TERRESTRIAL ISOPOD BIOLOGY

CRUSTACEAN ISSUES

9

General editor:
FREDERICK R. SCHRAM
Instituut voor Systematiek en Populatiebiologie, Amsterdam

A.A. BALKEMA / ROTTERDAM / BROOKFIELD / 1995

TERRESTRIAL ISOPOD BIOLOGY

Edited by
M.A. ALIKHAN
Laurentian University, Sudbury, Ontario

A.A. BALKEMA / ROTTERDAM / BROOKFIELD/ 1995

Published by
A.A. Balkema, P.O. Box 1675, 3000 BR Rotterdam, Netherlands (Fax: +31/10.4135947)
A.A. Balkema Publishers, Old Post Road, Brookfield, VT 05036, USA (Fax: 802.276.3837)

ISBN: 978-9-0541019-32

Printed in the Netherlands

Table of contents

Preface

In 1967, Dr. Dorothy E. Bliss of the American Museum of Natural History at New York, brought together for the first time at the New York meeting of the American Association for the Advancement of Science scientists with an interest in terrestrial isopods. Since then, three international symposia dealing with the 'Biology of Isopods,' the most successful of the land colonizing crustaceans, have been held in London (United Kingdom), Urbino (Italy), and Poitiers (France). The London meeting was sponsored by the Zoological Society of London, and its proceedings containing 21 papers, were edited by Stephen Sutton and David Holdich (Zoological Society of London Symposia 53, 1984). Proceedings of the second symposium, sponsored by Consiglio Nazionale delle Ricerche Unione Zoologica Italiana, containing 26 presentations, were published by Franco Ferrara, Roberto Argano, Claudio Manicastri, Helmut Schmalfuss, and Stefano Taiti [Monitore zoologico italiano (N.S.) Monografia 4, 1989]. The third meeting was financed by 'Faculté des Sciences de l'Université de Poitiers,' 'Conseil scientifique de l'Université de Poitiers,' 'Conseil Général de la Vienne' and 'Muncipalité de la Ville de Poitiers, France.' The volume, containing 26 papers and 12 posters presented during the meeting, was organized by Pierre Juchault and Jean Pierre Mocquard.

The present symposium was proposed for the 1992 meeting of the American Society of Zoologists by Professor Michael R. Warburg of Technion, Israel Institute of Technology at Haifa, during his stay during 1991 with me at the Laurentian University. This explains my involvement in the organization of the meeting, and the editing of its proceedings. The sponsorship was provided by The Crustacean Society. Unfortunately, because of the timing of the meeting, fewer than 20 colleagues from various parts of the world were able to attend. Nevertheless, more than twelve papers and four posters on various aspects of the biology of isopods were presented both during the symposium and the general session. Out of these, ten papers are being published in this volume in addition to some other contributions on the subject that were not part of the original symposium. The paper presented by Barbara Taylor and T. H. Carefoot on the 'Evolution of terrestrial life in isopods: loss of gas exchange & survival ability in water,' has been accepted for publication in the Canadian Journal of Zoology (1993).

I am grateful to Drs. Robert W. Elner and Richard Brusca of The Crustacean Society for their continued encouragement, to Dr. Thomas H. Carefoot of the Department of Zoology of the University of British Columbia, Dr. Cliff Crawford of the Biology Department of the University of New Mexico, and Dr. Michael Warburg of the Department of Biology at the Technion at Haifa, for their support and help in the organization of the symposium, to Dr. Stephano Taiti of Centro di Studio per la Faunistica ed Ecologia Tropicali, CNR, Florence (Italy) for providing me with the addresses of various scientists and sending out the notice of the meeting, to the American Society of Zoologists for the waiver of registration fee for the participants in the symposium, and to my various colleagues, who, because of the holiday season made personal sacrifice and attended the meeting. Last of all my thanks to Bridget Farley of the American Society of Zoologists for her support, advice and assistance. She deserves a part of the credit for the success of this symposium.

M.A. Alikhan
Sudbury, Canada
June 11, 1993

Introduction

Last time isopods were discussed on American soil was 25 years ago at the symposium on 'The terrestrial adaptations of Crustacea,' arranged by Dorothy Bliss and Eric Edney during the AAAS meeting in New York, December 1967. Meanwhile the isopodologists held three symposia in Europe where most of the research on this group was conducted. All of the symposia volumes have been published. One aim of this meeting was to stimulate research on American isopods some of which are endemic species and very little is known about them.

This fascinating group is the only crustacean group with representatives in all terrestrial ecosystems ranging from the sea shore to the desert. Several species have colonized deserts, some of them are highly successful and abundant. Thus, this group can be considered as a model for the successful transition on land.

Much of the early work was descriptive concerning structure. Other early work dealt with ecological and zoogeographical aspects. More recent research dealt with physiological adaptations related to water and thermal balance. In recent years, more attention was devoted to ecological studies dealing with reproductive strategies, and population and community structure.

The symposium deals with two main subjects: 1) The effects of stressful conditions on the individual animal as reflected by its survival, or by the disruption of its normal reproductive patterns. 2) The distribution of the isopods and their selection of microhabitat. Thus, one paper by Strus, Drobne, and Licar dealt with the comparative anatomy of the digestive system in two amphibious and two terrestrial isopod species, while another paper by Alikhan discusses the effect of metal pollutants on the metabolism. A third paper by Wright and O'Donnell describes the ways ammonia is excreted by a terrestrial isopod inhabiting (*Porcellio*) a mesic habitat. Very little is known about excretion in oniscid isopods, in particular of interest are the desert species because of their water shortage problem. Another paper by Heinzelmann, Crawford, Warburg, and Molles dealt with the effects of temperature and photoperiodism on the reproductive pattern in two isopod species: *Armadillo officinalis* and *Schizidium tiberianum*, while Dangerfield and Telford discussed the importance of life history variation with regards to reproductive tactic. The second subject was discussed in three lectures. One paper by Szlávecz dealt with the distribution of isopod species in an old park in Europe. Warburg and Weinstein dealt with the distribution of isopods as related to stone coverage in a Mediterranean grassland. Finally, a third paper by Hornung and Warburg discussed the correlation between vegetation and abundance of a cosmopolitan species, *Armadillidium vulgare*, in the southwestern USA. This interesting species is unique in its colonizing capabilities, and it is of great significance to study the causes for this successful colonization.

M.R. Warburg
Haifa, Israel

Comparative study of metal bioaccumulation, and oxygen and nitrogen metabolism in three terrestrial isopod species from contaminated and uncontaminated sites in Northeastern Ontario

M.A. Alikhan
Department of Biology, Laurentian University, Sudbury, Ontario, Canada

ABSTRACT

Copper and nickel concentration in whole *Metaponorthus pruinosus, Oniscus asellus* and *Porcellio spinicornis* (Isopoda, Crustacea), procured from metal-contaminated and uncontaminated sites in Northeastern Ontario (Canada), were related to the background levels in the habitat. Isopod species from contaminated sites had significantly higher metal concentrations in their body tissues than did those procured from uncontaminated sites. Difference in body tissue metal concentrations in various species were the function of the habitat and the body weight. Oxygen consumption rates, hepatopancreatic cell metabolic activity and plasma membrane surface potential in individuals procured from metal-contaminated sites were significantly higher than those of isopods obtained from uncontaminated sites. The significance of these findings is discussed.

1 INTRODUCTION

The degree of deterioration of natural ecosystems due to metal contamination, as well as the long-term effects of metal contaminants on biological systems, are difficult to assess by simple acute toxicity tests. Environmental concentrations of both class A and class B, as well as related metals, even in industrial barrens, are relatively lower than their LC_{50} values normally used as indices of susceptibility of organisms to these contaminants. The induction of physiological modifications by sublethal doses, however, may provide more realistic information on the stress these pollutants exert on various animals (Warren 1971, Waldichuck 1979). The physiological modifications, whenever they occur, disturb the energetic equilibrium in exposed organisms, and, as a consequence, may have both long- and short-term effects on the stability of their populations (Laughlin & Neff 1981). Such imbalances may arise from decreased nutritional energy because of relatively lower assimilation capabilities, or from increased respiratory requirements as a result of physiological stress. The physiological modifications in almost all instances are closely linked to the dysfunctioning of the enzyme systems at the molecular level (Bryan 1971).

The study of the energetic, electrical activity, and the oxygen requirements of various body tissues in animals not only indicate their prevailing metabolic rates, but they also provide a measure of the effects of both exogenous (= environmental) and endogenous (= physiological) parameters. Studies on both aquatic and terrestrial crustacean species have shown that the rates of oxygen consumption in these animals are not only influenced by extrinsic environmental parameters, such as seasons, latitude, temperature, photoperiod, strength of the geomagnetic field and the oxygen concentration of the respiratory mixture (Bridges & Brand 1980, Martel & Alikhan 1982, Alikhan 1983), but also by intrinsic physiological components, such as body weight, levels of activity, starvation, moult-stage, growth-stage, and biological circadian rhythms (Wolvekamp & Waterman 1960, Wycliffe & Job 1977, Bagatto & Alikhan 1986).

Earlier literature on the correlation of physical environmental factors with oxygen consumption and metabolic rates in both aquatic and terrestrial crustacean species has been reviewed by Wolvekamp & Waterman (1960), Lockwood (1967), Vernberg & Vernberg (1972), Newell (1973), and Prosser (1973). Recent studies on the rates of oxygen uptake from an encompassing range of crustacean types exposed to a variety of environmental conditions are discussed by McMahon & Wilkens (1983). Information on the relationship of the electrical activity of the cells with the prevailing metabolic rates under both uncontaminated and metal-contaminated conditions is provided by Alikhan & Naich (1987), Alikhan & Storch (1990), and Alikhan et al. (1985).

In previous studies, Alikhan (1993) provided information on the differentiation in copper and nickel accumulations in adult female *Porcellio spinicornis* Say, 1818 (Porcellionidae, Isopoda, Crustacea) obtained from both metal-contaminated and uncontaminated sites in Northeastern Ontario, and their progeny. The objective of the present study was to compare tissue energetics, as well as oxygen consumption and nitrogen excretion rates in three terrestrial isopod species procured from both copper and nickel contaminated and uncontaminated sites in Northeastern Ontario, Canada. Copper, a class B, sulphur- or nitrogen-seeking, element (Nieboer & Richardson 1980), is an integral part of the crustacean respiratory protein, haemocyanin, and it is accumulated within the cytoplasmic lysosomal vesicles (= 'cuprosomes' of Wieser 1968) of the hepatopancreas, the main regulatory organ in crustacean species (Alikhan & Storch 1990). Nickel, a borderline metal between class A (oxygen-seeking) and class B elements (Nieboer & Richardson 1980), on the other hand, is required by isopod species to sustain both muscular and neural action potentials (Simkiss & Taylor 1989). Survival of the freshwater isopod *Asellus aquaticus* (Latreille) is decreased with increasing dietary nickel concentrations, while low Ni levels in the diet of the terrestrial isopod *Oniscus asellus* adversely affect the regulatory ability of its hepatopancreas (Alikhan & Storch 1990). The data on the hepatopancreatic catalytic activity in the pyrogallol-oxidation chemiluminescence reaction provides information on the physiological state of the tissue, while the study of the surface potential provides information on the electrical activity of plasma membrane.

2 MATERIALS AND METHODS

2.1 *Animals*

Adult seventh growth-stage *Metoponorthus (Metoponorthus) pruinosus* (Brandt, 1833) (Porcellionidae, Isopoda), *Oniscus asellus* Linn, 1758 (Oniscidae, Isopoda), and *Porcellio spinicornis* Say, 1818 (Porcellionidae, Isopoda), males and females, were collected during May to July 1992 from wooded areas near the town of Copper Cliff (46 26' N 81 07' W), Ontario, Canada, and the 'Cup and Saucer Mountain' on Manitoulin Island near the town of Little Current (45 58' N 81 55' W), Ontario, Canada. Manitoulin Island, approximately 100 km southwest of Sudbury, is an uncontaminated site, whereas Copper Cliff sites are approximately 2 km downwind of the primary smelting works of the International Nickel Company of Canada, and are heavily contaminated with cadmium, copper, nickel, manganese and zinc (Alikhan et al. 1990). Growth-stage of the isopod was determined by the criterion of Alikhan (1972). Isopods from each population were brought to the laboratory and acclimated to an ambient temperature of 22° ± 2°C at 86% to 96% relative humidity, and a light (3.2 klx) : darkness cycle of 15L (0600-2100) : 9D (summer conditions in Northeastern Ontario). All isopods were caged in individual culture jars lined with wet gypsum and containing leaf litter and soil (upper 5-7 cm) from their 'own' site. Leaf litter residues were removed each week and weighed to the nearest 0.001 mg to determine weekly food consumption. To avoid fungal growth, faecal pellets were removed daily. As hepatopancreatic metal concentration differences between the two sexes were not significant at the 5% level, pooled hepatopancreatic tissue from either ten adult intermoult females or males of each species from each population were used for pyrogallol-oxidation chemiluminescence reaction, cell surface potential measurements and copper and nickel analysis.

2.2 *Atomic absorption spectrophotometry*

For copper and nickel concentration determinations, leaf litter and isopod hepatopancreas were placed in glass petri dishes and oven dried at 60°C (± 5°C) for 48 hours. After cooling to room temperature, samples were weighed to nearest 0.001 mg, digested in boiling concentrated aqua regia (3 mL Merck's 'suprapur' 65% nitric acid : 1 mL BDH analytical grade concentrated hydrochloric acid), diluted to 20 mL with 1*M* nitric acid, and analyzed for copper and nickel with a Perkin-Elmer atomic absorption spectrophotometer by the flame method. The detection limits of the atomic absorption spectrophotometer used for copper and nickel amounted to 0.077 and 0.15 $\mu g\ mL^{-1}$, respectively. Lobster copper and nickel standards were supplied by the National Research Council of Canada (Nrc) laboratories at Ottawa, Ontario, Canada. 'Metal coefficient' (= 'concentration factor' of Hopkin 1980) in whole isopods was calculated by dividing the concentration of the metal in the isopod by its levels in the leaf litter.

2.3 *Hypatopancreatic cell catalytical activity and plasma membrane potential*

Hepatopancreatic tubules were homogenized in a hand-operated disposable glass microtissue grinder, and their catalytic activity was determined by the modified pyrogallol-oxidation chemiluminscence technique of Alikhan et al. (1985). The cell plasma membrane surface potential was measured by the powder-electrode technique of Alikhan & Naich (1987). The chemiluminescence technique involved the use of a high voltage supply unit, an electronic scaler, and a photomultiplier with a photocathode spectral sensitivity range of 300-600 nm. The reaction mixture (without additional scintillation agents) consisted of 1.0 mL phosphate buffer (pH 6.8), 1.0 mL 1% hydrogen peroxide, 0.1 mL 1% pyrogallol (Sigma Chemical Co., St. Louise, MO), and 0.02 mL hypatopancreatic tissue homogenate in phosphate buffer. The mixture was poured into a quartz vial and the kinetics of the reaction (photon emission recorded as impulse frequency) were measured at a constant voltage of 1200 V.

2.4 *Oxygen consumption*

The oxygen consumption rates of randomly selected twenty-two seventh growth-stage individuals (11 males and 11 females) of the three species from each population was measured under laboratory acclimation conditions on day 21 after their transfer to the laboratory. Each animal was weighed, caged in a 5.0 mL respirometer flask of a Gilson Differential Respirometer and allowed to equilibrate for two hours before first of the three readings on its oxygen uptake was taken. The side-arm of the reaction flask contained 0.3 mL 20% potassium hydroxide solution for carbon dioxide absorption, and its central well held 0.3 mL 1N sulphuric acid for ammonia absorption. A piece of moist filter paper at the bottom of the flask counteracted the dehydrating effects of both potassium hydroxide and sulphuric acid. The isopods were not touched by hand during these manipulations. All experiments were conducted between 1030 and 1400 hours, and oxygen consumption in each case was measured at least for two hours, with readings being taken at 20-minute intervals. Eight such measurements were taken on each isopod, with their mean value ± standard error (S.E.) representing the oxygen uptake rate for that individual. Replicate values usually varied by less than 5%. Occasionally, isopods were very active in the reaction flask and, rarely, were found either dead or in the early stages of their posterior moulting. When this happened, the data on these animal were discarded and the experiment was repeated with other individuals of the same sex and approximately of the same weight from the same species and population. Regression equations of log oxygen uptake per hour per individual against log wet weight for the two populations were compared for differences in slope and elevation using an application of Student's *t* tests as described by Zar (1974). Overall data were analyzed using two- and three-way analyses of variance (ANOVA), combined with Tukey's Multiple Comparison Tests (Tukey's MCT; a = 0.05). Non-parametric tests were used to analyze data on growth and feeding rates when variance was found to be non-homogenous.

3 RESULTS

The data on copper and nickel concentrations in whole animals procured from the Manitoulin Island (uncontaminated) and the Copper Cliff (contaminated) sites are presented in Figure 1, whereas information on metal 'concentration-coefficient' in the hepatopancreas of intermoult isopods from the two sites is provided in Figure 2.

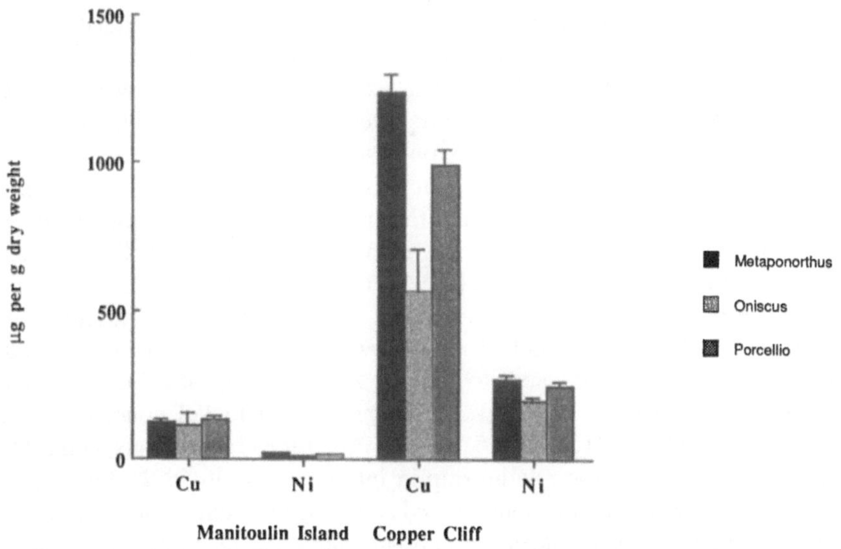

Figure 1. Copper and nickel concentrations in whole intermoult, seventh growth-stage males and females of three isopod species procured from Manitoulin Island (uncontaminated site) and Copper Cliff (contaminated site). Each point in the figure represents the mean value (± standard error) in 11 males and 11 females.

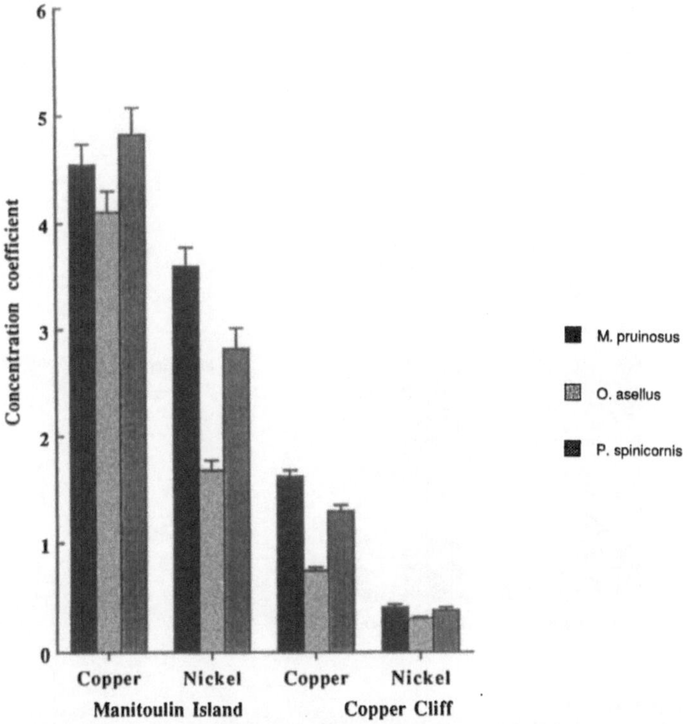

Figure 2. Metal 'concentration-coefficient' in the hepatopancreas of intermoult species, seventh growth-stage, males and females of three isopod species procured from the Manitoulin Island (uncontaminated site) and Copper Cliff (contaminated site). Each point in the figure represents the mean value (± standard error) in 11 males and 11 females.

Mean values (± S.E.) on hepatopancreatic catalytic activity and cell surface membrane potentials in intermoult adults are plotted in Figures 3 and 4, respectively, whereas the oxygen consumption and ammonia excretion rates (μL g^{-1} wet weight $hour^{-1}$) in the three species from both sites are reported in Table 1. The relationship of isopod wet live-weight and oxygen consumption is summarized in Figure 5.

Mean copper concentrations in whole individuals of *M. pruinosis, O. asellus,* and *P. spinicornis* procured from uncontaminated (Manitoulin Island) sites amounted to 125.6 ± 20.2, 115.0 ± 5.2 and 135. 7 ± 7.5 μg. g^{-1} wet weight, respectively, whereas mean nickel concentrations amounted to 21.3 ± 1.5, 9.9 ± 1.0 and 16.7 ± 0.9 μg·g^{-1} wet weight, respectively (Fig. 1). In comparison, mean copper concentrations in whole *M. pruinosis, O. asellus,* and *P. spinicornis* trapped from contaminated (Copper Cliff) sites were calculated to be 1, 237.9 ± 59.3, 567.3 ± 27.4 and 989.0 ± 84.6 μg·g^{-1} wet weight, respectively, whereas mean nickel concentrations amounted to 273.3 ± 11.8, 201.1 ± 10.5 μg·g^{-1} wet weight, respectively (Fig. 1). Whole body copper concentration relationship in intermoult adults in the three species from both uncontaminated and contaminated sites was *O. asellus* < *M. pruinosus* < *P. spinicornis*, whereas nickel concentration relationship was *O. asellus* < *P. spinicornis* < *M. pruinosus* (Fig. 1). Difference between the two sexes was not significant

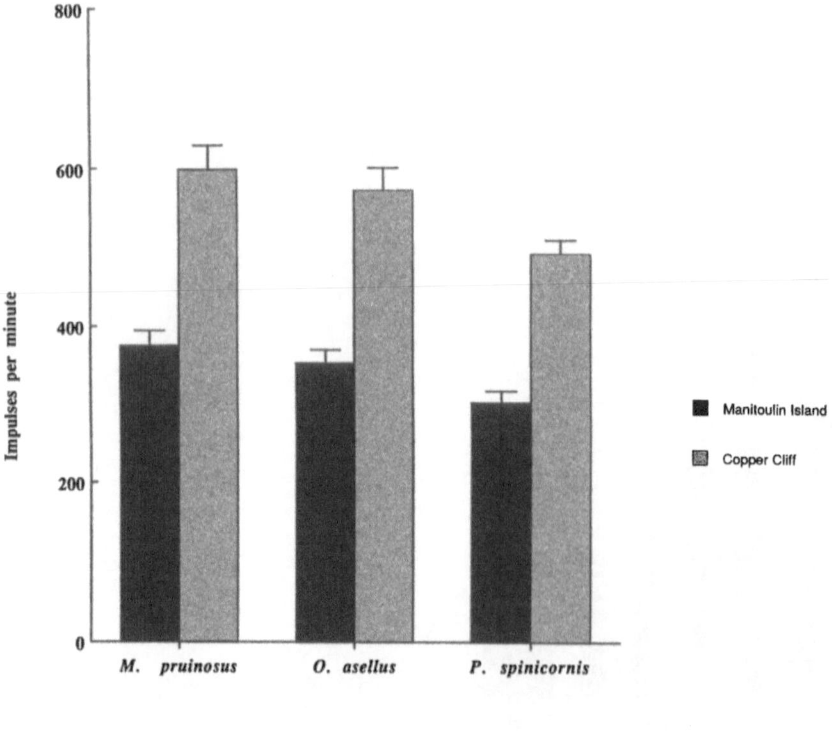

Figure 3. Mean catalytic activity (± standard error) of the hepatopancreas of intermoult species, seventh growth-stage, males and females of three isopod species procured from the Manitoulin Island (uncontaminated site) and Copper Cliff (contaminated site). Each point in the figure represents the mean value (± standard error) in 11 males and 11 females.

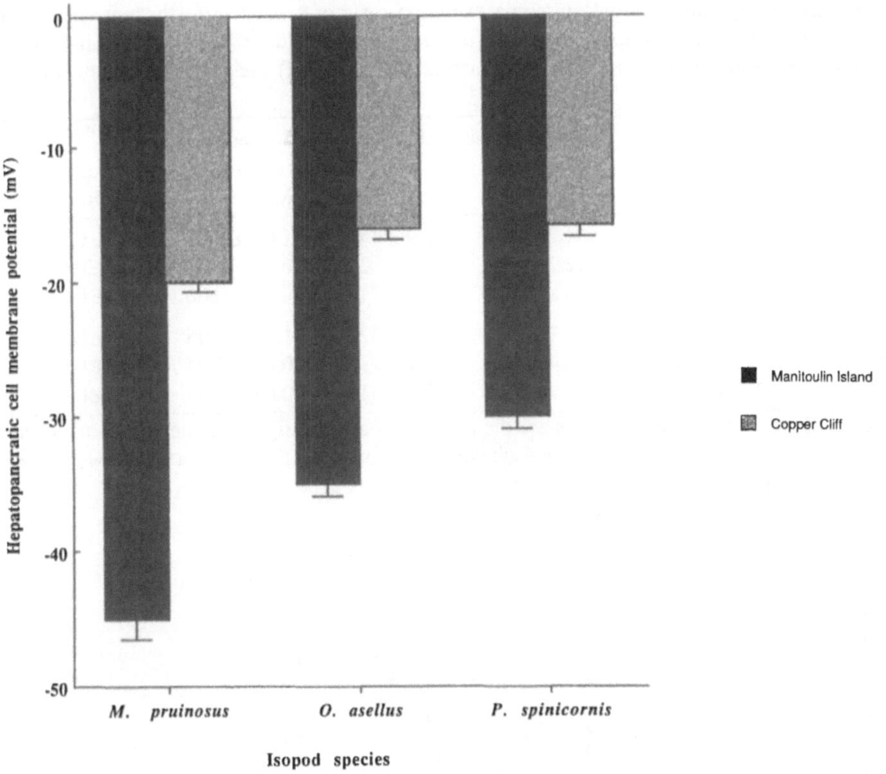

Figure 4. Mean cell surface electrical potential (± standard error) of the hepatopancreas of intermoult species, seventh growth-stage, males and females of three isopod species procured from the Manitoulin Island (uncontaminated site) and Copper Cliff (contaminated site). Each point in the figure represents the mean value (± standard error) in 11 males and 11 females.

at the 5% level. Hepatopancreatic metal 'concentration-coefficients' were significantly ($P < 0.01$) higher in isopods procured from Manitoulin Island (uncontaminated site) than in those obtained from Copper Cliff (contaminated) sites (Fig. 2).

Hepatopancreatic mean catalytic activity (± S.E.) in *O. asellus*, *P. spinicornis* and *M. pruinosus* obtained from Manitoulin Island (uncontaminated site) amounted to 352 ± 17.6, 302 ± 15.1 and 375 ± 18.8 impulses min^{-1}, respectively, whereas in isopods trapped from Copper Cliff it was calculated to be 491.3 ± 16.1, 570.0 ± 28.5 and 596.5 ± 30.1 impulses min^{-1}, repectively (Fig. 3). Similarly, cell membrane surface potential of hepatopancreatic cells from animals obtained from Copper Cliff sites were twice as much as (in terms of absolute values) those of the hepatopancreas obtained from isopods procured from Manitoulin Island (Fig. 4).

Both oxygen consumption and ammonia excretion rates, as may be observed from the data reported in Table 1 and Figure 5, were significantly influenced ($P < 0.05$) by the sampling site and the species of the isopod. Thus, isopods procured from contaminated sites consumed oxygen at a significantly ($P < 0.05$) higher rate than did those obtained from uncontaminated site (Table 1). Differences in oxygen uptake due to sex of the isopod, except for that in *O. asellus* procured from contaminated sites,

Table 1. Oxygen consumption and ammonia excretion rates of seventh growth-stage intermoult adults of *Metaponorthus pruinosus*, *Oniscus asellus* and *Porcellio spinicornis* obtained from uncontaminated (Manitoulin Island) and contaminated (Copper Cliff) sites in Northeastern Ontario, Canada.

Collection site	Species	Oxygen consumption (μL g^{-1} wet wt hr^{-1} ± S.E.*)		Ammonia excretion (μL N g^{-1} wet wt hr^{-1} ± S.E.*)	
		Male	Female	Male	Female
Manitoulin island	*M. pruinosus*	464.2 ± 5.5	485.9 ± 14.3	1.07 ± 0.01	1.05 ± 0.05
	O. asellus	442.7 ± 14.2	465.6 ± 2.1	0.61 ± 0.02	0.89 ± 0.06
	P. spinicornis	326 ± 12.8	443.2 ± 21.8	0.60 ± 0.06	0.61 ± 0.15
Copper cliff	*M. pruinosus*	571.1 ± 13.5	605.7 ± 32.3	1.09 ± 0.05	1.09 ± 0.07
	O. asellus	552.4 ± 18.3	598.7 ± 28.7	0.87 ± 0.04	0.91 ± 0.05
	P. spinicornis	501.9 ± 22.3	572.3 ± 31.9	0.65 ± 0.05	0.72 ± 0.09

*Standard error of the mean. Values are the average of at least eight readings on 11 males and 11 females of each species. Means within each column followed by the same letter and within each row followed by the same figure are not significantly different at the 5% level.

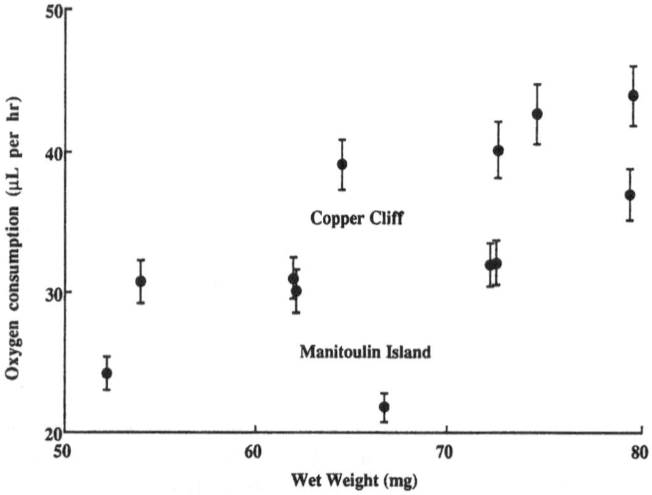

Figure 5. The relationship of oxygen consumption and the wet live weight of the isopod species procured from the Manitoulin Island (uncontaminated site) and Copper Cliff (contaminated site). Each point in the figure represents the mean value (± standard error) in 11 males and 11 females.

were significant at $P < 0.05$ (Table 1). Differences between the oxygen uptake rates in *O. asellus* males and females were not significant at $P > 0.05$ (Table 1). In general, oxygen consumption rate in each experimental isopod was related to its wet weight; the heavier the animal was, the lesser was its oxygen uptake rate (Fig. 5). However, in spite of individual differences, over all oxygen consumption rates in the three species from the uncontaminated and the contaminated sites were *P. spinicornis* < *M. pruinosus* < *O. asellus* and *M. pruinosus* < *P. spinicornis* < *O. asellus* (Fig. 5) respectively.

Mean ammonia production ranged from 0.60 ± 0.06 μL N g^{-1} wet weight hr^{-1} in male *P. spinicornis* from Manitoulin Island to 1.09 ± 0.07 μL N g^{-1} wet weight hr^{-1} in female *M. pruinosus* from Copper Cliff (Table 1). However, differences due to site, sex and species were not significant at the 5% level (Table 1).

4 DISCUSSION

In agreement with the observations of Wieser & Makart (1961), Hopkin & Martin (1982) and Alikhan & Storch (1990), both males and females of *M. pruinosus, O. asellus,* and *P. spinicornis*, procured from the two sites (uncontaminated Manitoulin Island and contaminated Copper Cliff sites), contained both Cu and Ni in their body-tissues. 'Bioconcentration coefficient' (= 'concentration factor' of Hopkin 1989) of these two metals, in agreement with the findings of Hopkin & Martin (1982), were significantly higher in animals procured from Manitoulin Island (uncontaminated site) than in isopods collected from Copper Cliff sites. This would suggest that isopods from contaminated sites assimilated significantly lower amounts of copper and Ni than did animals from the uncontaminated site. This limitation of net assimilation of the two metals by the isopods from the contaminated sites could be due to either saturation of cellular metal uptake mechanism or an active increase in the metal-excretion rate. The later suggestion, however, was questioned by Hopkin & Martin (1984) who suggested that retention rather than excretion of metals from within the cells was more economical for the isopod because it expanded less energy than would be needed for their excretion.

According to Alikhan (1992), lower metal 'concentration-coefficient' at relatively higher background levels might also be attributable to the tolerance for higher tissue metal concentrations in isopod populations inhabiting contaminated sites. Nevertheless, higher 'bioconcentration coefficient' in isopods at relatively low metal background levels have as yet not been clearly demonstrated, although it has been suggested that such phenomenon might occur with Cd (Hopkin et al. 1986). Recently, Bardeggia & Alikhan (1991) have demonstrated that Cu accumulation by whole crayfish is retarded by the presence of approximately 100 μg.g^{-1} of Ni in the diet, whereas increasing Ni to 500-600 μg. g^{-1} may enhance it. Therefore, it appears that relatively low metal 'bioconcentration coefficients' could be the product of all of the above mentioned phenomena.

Oxygen consumption in isopods is known to be affected by both environmental parameters, such as ambient temperature (Husain & Alikhan 1979, Newell & Bayne 1973, Wieser 1972), photoperiod (Newell et al. 1976, Wieser et al. 1969), environmental oxygen concentration (Husain & Alikhan 1979), and magnetic field intensity (Martel & Alikhan 1982), and endogenous factors, such as body weight (Bukhari & Alikhan 1984, Hemmingsen 1960, Husain & Alikhan 1979), growth-stage (Bagatto & Alikhan 1986), and the nutritional status (Bagatto & Alikhan 1986, Marsden et al. 1973). Rhythmic changes in the oxygen uptake rates (Aldrich 1975), however, sometimes complicate the interpretation of the influence of both environmental and endogenous factors. Changes in oxygen respiration rates in the terrestrial isopod, *Porcellio scaber* Latreille, 1804 (Porcellionidae, Isopoda) and in the estuarine mysid, *Leptomysis lingvura* G.O. Sars (Mysidacea, Crustacea) after metal exposure, are discussed by Joosse & van Vliet (1984) and Gaudy et al. (1991), respectively.

High oxygen consumption values due to excitation because of handling is well known among both vertebrate (Saunders 1963) and invertebrate (including sessile and sedentary marine) species (Thompson & Byne 1972, Gaffney & Diehl 1986). Such changes, according to Jobing (1981), maybe minimized by careful acclimation procedures prior to respiration rate determinations. According to Newell et al. (1974), Vo_2 in the isopod *Porcellio scaber* an increase by 48% through handling, change in light intensity and contact with food-an increase which is not related to locomotory activity. Handling and other effects were minimized in the present study by allowing the animals to acclimate to laboratory conditions for 21 days and get used to the respirometer flask for at least two hours. Even then it is expected that oxygen consumption rates in experimental animals may have been elevated.

Apart from size-related metabolic rates, oxygen consumption rates observed in the present study were significantly higher in isopods procured from the metal-contaminated sites than those from the uncontaminated habitat. This is to be expected as copper, even in concentrations amounting to 0.1 mg L^{-1}, is known to affect directly the neural centres in pinfish *Lagodon rhomboides* that are responsible for temporal variability of locomotor activity (Scarfe et al. 1982), and hence increased oxygen consumption rates. Such data, however, are limited and may not allow generalization to natural populations.

It is possible that elevated oxygen consumption rates, and hence the increased cell metabolism, as shown by the relatively higher irritability (= plasma membrane surface potential) and catalytic activity of the hepatopancreatic cells, is a function of genetic resistance in animals to high metal levels of their metal-contaminated environment, since in short-term adjustment (= physiological tolerance) to the presence of sublethal amounts of metals in the diet or the environment, the oxygen consumption rates in isopods, as reported by Alikhan & Pani (1989), show a significant decrease (while ammonia excretion may undergo an increases). Similar changes in metabolic rates in *Tubifex* exposed for short periods (approximately six weeks) to cadmium, mercury, zinc, chromium, and nickel (Brakovic-Popovic & Popovic 1977) and in Collembola fed for a month and a half on lead-enriched diets (Joosse & Verhoff 1983) have been attributed to a lower food intake.

The evolution of genetic resistance to metals is well documented for plants (Antonovics et al. 1971, Klerks & Weis 1987, Macnair 1987), but the phenomenon has as yet not been demonstrated in natural animal populations. The presence of elevated metal concentrations in tissues of several vertebrate and invertebrate species, especially in aquatic animals from contaminated areas, have been well documented (Alikhan et al. 1990, Bryan 1971), but the data on metal resistance in aquatic isopods attributable to genetic differences among populations inhabiting areas with different metal background levels have been presented only by Brown (1976). Recently, Alikhan (1993), on the basis of his laboratory studies on female *P. spinicornis* procured from contaminated and uncontaminated sites in Northeastern Ontario, has suggested the possibility of genetic differences between isopod populations from the two habitats controlling, at least partially, the hepatopancreatic bioaccumulation of calcium, magnesium and nickel. Obviously because of genetic differences, significant differences in physiological parameters in individuals of the same species procured from metal-contaminated and uncontaminated areas can be expected.

According to Prosser (1973), adaptations in animals to their environment are attributable to the interaction among several responses that guarantee the survival of the species. In Crustacea, adaptation to a particular environment seems to be related to physiological, morphological and ethological responses that vary in degree according to their metabolic level, thickness of the exoskeleton and activity as has been recorded among others by Rosas et al. (1992). As hepatopancreas in crustacean species is the main regulatory organ, as such it must achieve a delicate balance between various responses to ensure the survival of the species. The close relationship between the metabolic rate and the metal levels in the environment (both internal and external) may well be the characteristic of both aquatic and terrestrial crustacean species. Presently studies are in progress to understand the mystery of the development of metal tolerance in terrestrial and aquatic crustacean species, and to sort out the genetic variation in the individuals of the same species occupying metal-contaminated and uncontaminated habitats.

ACKOWLEDGEMENTS

The study was supported through a Salary Research Grant from the Laurentian University. Additional funding was provided by the Ontario Ministry of Northern Affairs and Mines through its NorTop programme.

REFERENCES

Aldrich , J. C. 1975. On the oxygen consumption of the crabs *Cancer pagurus* (L.) and *Maia squinado* (Herbst.). *Comp. Biochem. Physiol.* 50A:223-228.

Alikhan, M. A. 1972. Changes in the hepatopancreatic metabolic reserves of *Porcellio laevis* Latreille during starvation and the moult-cycle. *Amer. Midl. Natur.* 87: 503-514.

Alikhan, M. A. 1983. Oxygen consumption and haemolymph pressure measurements in *Armadillidium vulgare* (Latreille) (Armadillidiidae:Isopoda) during ecdysis. *J. Crust. Biol.* 3: 25-33.

Alikhan, M. A. 1993. Differentiation in copper and nickel accumulation in adult female and juvenile *Porcellio spinicornis* from contaminated and uncontaminated sites in Northeastern Ontario. *Bull. Environ. Contam. Toxicol.* 50: 922-928 .

Alikhan, M. A. & M. Naich 1987. Changes in counts of haemocytes and in their physicochemical properties during the moult cycle in *Porcellio spinicornis* Say (Porcellionidae, Isopoda). *Can. J. Zool.* 65: 1685-1688.

Alikhan, M. A. & A. K. Pani 1989. Magnesium and manganese uptake and accumulation in the body tissue and their effect on the oxygen and nitrogen metabolism in *Porcellio spinicornis* Say, 1818 (Porcellionidae : Isopoda). *Monitore Zool. Ital. (N. S.) Monogr.* 4 : 211-225.

Alikhan, M. A. & V. Storch 1990. Copper and nickel uptake and accumulation, and their effects on redox and electrical potentials of hepatopancreatic cells of *Oniscus asellus* Linnaeus (Porcellionidae, Isopoda). *Can. J. Zool.* 68: 651-655.

Alikhan, M. A., G. Bagatto & S. Zia 1990. The crayfish as a 'biological indicator' of aquatic contamination by heavy metals. *Wat. Res.* 24: 1069-1076.

Alikhan, M. A., A. Bedarek & S. Grabiec 1985. The physiological and morphological characteristics of *Neoaplectana carpocapsae* (Nematoda, Steinernematidae) in two insect hosts. *J. Invert. Pathol.* 45: 168-173.

Antonovics, J., A. D. Bradshaw & R. G. Turner 1971. Heavy metal tolerance in plants. *Adv. Ecol. Res.* 7: 1-85.

Bagatto, G. & M. A. Alikhan 1986. Effect of growth-stage and nutritional status on the oxygen consumption and ammonia production in *Porcellio spinicornis* Say (Isopoda, Porcellionidae). *Crustaceana* 51: 1-8.

Bardeggia, M. & M. A. Alikhan 1991. The relationship between copper and nickel levels in the diet, and their uptake and accumulation by *Cambarus bartoni* (Fabricius) (Decapoda, Crustacea). *Wat. Res.* 25: 1187-1192.

Brakovic-Popovic, I & M. Popovic 1977. Effects of heavy metals on survival and respiration rate of Tubificid worms: part II.-Effects of respiration rate. *Environ. Pollut.* 13: 93-98.

Bridges, C.R. & A. R. Brand 1980. The effect of hypoxia on oxygen consumption and blood lactate levels of some marine Crustacea. *Comp. Biochem. Physiol.* 65A: 457-462.

Brown, B. E. 1976. Observation on the tolerance of the isopod *Asellus meridianus*. Reac. to copper and lead. *Wat. Res.* 10: 555-559.

Bryan, G. W. 1971. The effect of heavy metals (other than mercury) on marine and estuarine organisms. *Proc. R. Soc. Lond.* (B) 177: 389-410.

Gaffney, P. M. & W. J. Diehl 1986. Growth, condition and specific dynamic action in the mussel *Mytilus edulis* recovering from starvation. *Mar. Biol.* 93: 401-409.

Gaudy, R., J.-P. Guérin & P. Kerambrun 1991. Sublethal effects of cadmium on respiratory metabolism, nutrition, excretion and hydrolase activity in *Leptomysis lingvura* (Crustacea: Mysidacea). *Mar. Biol.* 109: 493-501.

Hemmingsen , A. M. 1960. Energy metabolism as related to body size and respiratory surfaces and its evolution. *Rep. Steno Hosp. Copenhagen* 9: 7-110.

Hopkin, S. P. & M. H. Martin 1982. The distribution of zinc, cadmium, lead and copper within the woodlouse *Oniscus asellus* (Crustacea, Isopoda). *Oecologia, Berl.* 54: 227-232.

Hopkin, S. P., G. N. Hardisty & M. H. Martin 1986. The woodlouse *Porcellio scaber* as a 'biological indicator' of zinc, cadmium, lead and copper pollution. *Environ. Pollut.* (B) 11: 271-290.

Husain, M. Z. & M. A. Alikhan 1979. A comparative study of the respiratory metabolism in *Porcellio laevis* Lat. and *Armadillidium vulgare* (Lat.): response to temperature, photoperiod and oxygen concentration. *Arch. Intern. Physiol. Biochim.* 87: 697-710.

Jobing, M. 1981. The influence of feeding on the metabolic rate of fish: a short review. *J. Fish. Biol.* 18: 385-400.

Joosse, E. N. G. & S. C. Verhoff 1983. Lead tolerance in Collembola. *Pedobiologia* 25: 11-18.

Joosse, E. N. G. & L. H. H. van Vliet 1984. Iron, manganese and zinc input in soil and litter near a blast-furnace plant and the effects on respiration of woodlice. *Pedobiologia* 26: 249-255.

Klerks, P. L. & J. S. Weis 1987. Genetic adaptation to heavy metals in aquatic organisms: a review. *Environ. Pollut.* 45: 173-205.

Laughlin, R. B. & J. M. Neff 1981. Ontogeny of respiratory and growth responses of larval mud crab *Rhithropanoplus harrisii* exposed to different temperatures, salinities and naphthalene concentrations. *Mar. Ecol. Prog. Ser.* 15: 319-332.

Lockwood, A. P. M. 1967. *Aspects of the Physiology of Crustacea.* San Francisco: Freeman.

Macnair, M. R. 1987. Heavy metal tolerance in plants: a model evolutionary system. *Trends Ecol. Physiol.* 2: 354-359

Marsden, I. D., R. C. Newell & M. Ahsanullah 1973. The effect of starvation on the metabolism of the shore crab, *Cancer maenas. Comp. Biochem. Physiol.* 45A: 195-213.

Martel, M. R. & M. A. Alikhan 1982. The effect of magnetic field intensities on the oxygen consumption in *Porcellio laevis* Latre.(Oniscoidea, Isopoda). *Arch. Int. Physiol. Bichim.* 90: 111-122.

McMahon, B. R. & J. L. Wilkens 1983. Ventilation, perfusion, and oxygen uptake. In D. E. Bliss (ed.) *The Biology of Crustacea*, 5: 289-372. New York: Academic Press.

Newell, R. C. 1973. Factors affecting the respiration of intertidal invertebrates. *Amer. Zool.* 13: 513-528.

Newell, R. C. 1976. Adaptations to intertidal life. In R. C. Newell (ed.) *Adaptation to Environment: Studies on the Physiology of Marine Animals*: 112-134. London: Butterworth.

Newell, R. C. & B. L. Bayne 1973. A review on temperature and metabolic acclimation in intertidal marine invertebrates. *Neth. J. Sea Res.* 7: 421-433.

Newell, R. C., W. Wieser & V. I. Pye 1974. Factors affecting the oxygen consumption in the woodlouse *Porcellio scaber* Latr. *Oecologia* 16-31-51.

Nieboer, E. & D. H. S. Richardson 1980. The replacement of the nondescript term 'heavy metals' by a biologically and chemically significant classification of metal ions. *Environ. Pollut.* 1B: 3-26.

Prosser, C. L. 1973. *Comparative Animal Physiology*. Philadelphia: W. B. Saunders Co.

Rosas, C., A. Sanchez, E. Escobar, L. Soto & A. Bolongaro-Crevenna 1992. Daily variations of oxygen consumption and glucose hemolymph level related to morphological and ecological adaptations of Crustacea. *Comp. Biochem. Physiol.* 101A: 323-328.

Saunders, R. L. 1963. Respiration of the Atlantic cod. *J. Fish. Res. Bd. Can.* 20: 373-386.

Scarfe, A. D., K. A. Jones, C.W. Steele, H. Kleerekoper & M. Corbett 1982. Locomotor behavior of four marine teleosts in response to sublethal copper exposure. *Aquat. Toxicol.* 2: 335-353.

Simkiss, K. & M. G. Taylor 1989. Metal fluxes across the membranes of aquatic organisms. *CRC Crit. Rev. Aquat. Sci.* 1: 173-188.

Thompson, R. J. & B. L.Byne 1972. Active metabolism associated with feeding in the mussel *Mytilus edulis* L. *J. Exp. Mar. Biol. Ecol.* 9: 111-124.

Vernberg, W. B. & F. J. Vernberg 1972. *Environmental Physiology of Marine Animals.* New York: Springer-Verlag.

Waldichuck, M. 1979. Review of the problem. *Phil. Trans. R. Soc.* (B) 286: 399-424.

Warren, C. E. 1971. *Biology and Water Pollution Control.* Philadelphia: W. B. Saunders.

Wieser, W. 1968. Aspects of nutrition and the metabolism of copper in isopods. *Amer. Zool.* 8: 495-506.

Wieser, W. 1972. Oxygen consumption and ammonia excretion in *Ligia beaudiana* M. E. *Comp. Biochem. Physiol.* 43A: 869-876.

Wieser, W. & H. Makart 1961. Der Sauerstoff verbrauch und der Gehalt an Ca, Cu and einigen anderen Spurenelementen bei terrestrischen Asseln. *Z. Naturforsch.* 16b: 816-819.

Wieser, W., G. Schweizer & R. Hartenstein 1969. Patterns in the release of gaseous ammonia by terrestrial isopods. *Oecologia* 3: 390-400.

Wolvekamp, H. P. & T. H. Waterman 1960. Respiration. In T. H. Waterman (ed.) *Physiology of Crustacea* 1: 35-100. New York: Academic Press.

Wycliffe, M. J. & S. V. Job 1977. Standard, routine and active oxygen consumption of a freshwater shrimp. *Hydriobiologia* 54: 33-39.

Zar, J. H. 1974. *Biostatistical Analysis.* Philadelphia: Prentice-Hall.

Comparative anatomy and functional aspects of the digestive system in amphibious and terrestrial isopods (Isopoda: Oniscidea)

J. Štrus, D. Drobne & P. Ličar
Department of Biology, University of Ljubljana, Ljubljana, Slovenia

ABSTRACT

The functional morphology of the digestive system in two amphibious and two terrestrial isopod species was studied by means of light and electron microscopy. In *Ligia italica* and *Ligidium hypnorum* the digestive canal comprises straight ectodermal foregut and hindgut and a short endodermal midgut with midgut glands. The structural features of the foregut in these species demonstrate that its functions mainly are to press, filter and channel the food to the midgut glands and the hindgut. The hindgut is a simple tube with loose muscular network, smooth cuticular lining and without typhlosole. Its main functions are transport and absorption of nutrients and formation of faecal pellets. The digestive canal of *Trachelipus illyricus* and *Porcellio scaber* consists of ectodermal foregut and hindgut and two pairs of midgut glands caeca. The foregut in terrestrial isopods comprising a well-developed masticatory apparatus, tooth-like cuticular spines and large atrium, is more complex than in amphibious species. It mainly triturates and filters the food. The hindgut, containing a prominent typhlosole, thick muscular layers, diversity of cuticular spines and abundant microflora, is divided into functionally different anterior and papillate regions. The hindgut is an important passage-way for nutrients, where food is both mechanically and chemically transformed and separated into fractions. Nutrients are partly returned to the foregut via typhlosole channels and mixed with digestive juices in the large atrium. The basal parts of hindgut cells of papillate region bulge into haemolymph lacunae. Ultrastructural features of the cells suggest that they are involved in ion and water transport. The endodermal part of the digestive system is restricted to two pairs of midgut gland diverticula in higher oniscideans. In primitive amphibious species an endodermal midgut and paired midgut gland diverticula are present. Ultrastructural analysis of the midgut and midgut glands demonstrates that they are involved in the secretion of digestive enzymes, absorption and storage of lipids respectively.

1 INTRODUCTION

Oniscidea, a group of terrestrial isopods commonly called woodlice, comprises more than two thousand species with a world-wide distribution (Vandel 1960). Specific structural and physiological adaptations have enabled woodlice to invade habitats from the seashore to extremely dry environments, including the deserts. They have evolved different feeding strategies to cope with conditions encountered in diverse terrestrial environments. The functional morphology of the digestive system, therefore, reflects specific adaptations for water conservation and an effective assimilation of nutrients from the food. The structure of the digestive system has been analyzed in many isopod crustaceans (Goodrich 1939, Holdich & Mayes 1975, Tuzet et al. 1960, Jones 1968, Alikhan 1969, Vernon et al. 1974, Hassal & Jennings 1975, Ličar 1975, Wägele et al. 1981, Lane 1988, Hames & Hopkin 1989), but comparative studies on functional adaptations of the digestive system in primitive and higher oniscideans are relatively rare (Donadey 1973, Storch & Štrus 1989). The digestive system of isopods consists of a foregut (stomodeum) and hindgut (proctodeum), both lined with a cuticle, and an endodermal midgut (mesenteron). The endodermal part in most oniscideans consists of two pairs of midgut glands caeca. A short midgut situated between the foregut and the hindgut, connected to the midgut glands caeca, was described in some primitive species (Goodrich 1939, Holdich & Ratcliffe 1970, Štrus & Drašlar 1988). The microscopic anatomy of the ectodermal digestive tract and its role in mastication, transport, absorption and storage of food was investigated in several isopods (Storch & Štrus 1989, Hryniewiecka-Szyfter & Storch 1986) and its ability to recycle and retain digestive fluids was demonstrated recently in *Oniscus asellus* (Hames & Hopkin 1989).

A detailed ultrastructural analysis of the foregut in *Porcellio scaber* was performed by Storch (1987), who showed that the foregut was a complex part of the digestive tract where food was masticated, pressed, filtered or even degraded enzymatically. A comparative study of the isopod foregut and the phylogenetic implications of its structural features in different isopod groups were presented by Wägele (1989). The results of his study demonstrate that it is a valuable subject for a phylogenetic analysis of homologous structures and for the evaluation of structural adaptations to different feeding strategies. The posterior part of the foregut in higher oniscideans is surrounded by a circular fold of the anterior hindgut.

The anterior part of the hindgut with a dorsal typhlosole differs structuraly and functionally from the posterior papillate region. Absorption of nutrients and recycling of digestive fluids take place in the anterior part, while faeces is compacted in the posterior part which is also involved in transport and storage of water (Hames & Hopkin 1989, Warburg & Rosenberg 1989, Palackal et al.1984). The abundant microflora associated with the hindgut cuticle was observed in the anterior hindgut and typhlosole channels of several terrestrial isopods and its possible role in enzymatic degradation of food was discussed (Reyes & Tiedje 1976, Griffiths & Wood 1985, Hames & Hopkin 1989, Ulrich et al. 1991).

The midgut glands have been thoroughly studied in different isopods (Jones et al. 1969, Clifford & Witkus 1971, Donadey & Besse 1972, Hryniewiecka-Szyfter & Tyczewska 1978, Storch 1982, Bettica et al. 1984, Štrus 1987, Hames & Hopkin 1991). The monolayered digestive epithelium with two cell types plays an important role in food absorption, secretion, storage of lipids, metal accumulation, and excretion. The ultrastructure of large cells which store lipids is greatly affected by the dietary conditions of the animal (Storch 1982, 1984, Štrus et al. 1985, Štrus 1987).

The small cells with different types of intracellular granules were described as sites of metal accumulation and waste products excretion (Alikhan 1972, Wieser et al. 1976, Prosi et al. 1983, Hopkin 1986, Prosi and Dallinger 1988). The numerous studies on the digestive system of isopods demonstrate its large structural complexity. This paper presents a comparative study of the digestive system in two amphibious and two strictly terrestrial isopods. It was our aim to analyze the fine structure of the masticatory apparatus of the foregut, the foregut-hindgut junction, the muscular layer of the hindgut, the typhlosole channels and the luminal surface of the hindgut and to show how these structures may reflect specific feeding habits of the species.

2 MATERIALS AND METHODS

2.1 *Experimental animals*

Isopods were collected from their natural habitats in Slovenia. Specimens of *Ligia italica* Fabricius, 1798, were taken from the seashore in Piran (Gulf of Trieste), *Ligidium hypnorum*, Cuvier 1792, from the edge of the forest in the vicinity of Laško, *Trachelipus illyricus*, Verhoeff 1901, from the karstic area near Postojna, and *Porcellio scaber*, Latreille, 1804, from the stony area in Laško. Adult isopods of both sexes were brought to the laboratory and kept in plastic vials without food for several days. Segments of the digestive system were dissected out and fixed for light and electron microscopy.

2.2 *Light microscopy*

Different parts of the digestive tract were fixed in formol-ethanol and Carnoy B fixatives, dehydrated in ethanol series and embedded in paraplast. Serial cross and longitudinal sections were stained with hematoxylin-eosin or with the one step Azan technique for crustacean tissues (Hubschman 1962). Histological sections were examined and photographed in a Zeiss Axiophot light microscope.

2.3 *Scanning electron microscopy*

Isolated components of the digestive tract were fixed in a modified Karnowsky fixative (1% glutaraldehyde and 0.4% paraformaldehyde in 0.2 cacodylate buffer, pH = 7.2) for 2 hours at 4°C. Samples were postfixed in 1% osmium tetroxide in cacodylate buffer, followed by the TOTO procedure (Davies & Forge 1987). After dehydration through a graded series of ethanols, the samples were transferred to amylacetate and critical-point dried in liquid CO_2. Dried components of the digestive system were mounted on brass stubs, sputter-coated with gold, and examined in a Jeol 840A electron microscope.

2.4 *Transmission electron microscopy*

Components of the digestive system were dissected out and fixed in 3.5% glutaraldehyde in 0.1M cacodylate buffer, postfixed in 1% osmium ferrocyanide, stained en bloc with 1% uranyl acetate in maleate buffer overnight at 4°C, dehydrated in ethanols, and embedded in Spurr's medium (Spurr 1969). Ultrathin sections were examined in a Jeol 100S electron microscope.

3 RESULTS

3.1 *Foregut*

The gross morphology of the foregut is principally the same in all four species. Nevertheless, ultrastructural analysis of the main functional parts reveals differences in their structural complexity. The foreguts of *Ligia italica and Ligidium hypnorum* are flattened dorso-ventrally and have long ventral lamellae which extend deeply into the hindgut (Pl. 1A), while the foreguts of *Trachelipus illirycus* and *Porcellio scaber* are stout with short ventral lamellae (Pl. 1B).

The most prominent structures of the anterior foregut are lateralia which press and channel food particles to the posterior part of the foregut. The entire surface of the lateralia is covered with long spines in the two species of the family Ligiidae, with the feather-like spines situated predominantly on their ventral surfaces (Pl. 2). In *Ligidium hypnorum*, the ventral face of lateralia bears short spines surrounded with long pennate spines (Pl. 3A). The masticatory apparatus, as described in *Porcellio scaber* and *Oniscus asellus* (Storch 1987, Hames & Hopkin 1989) is not present in the foreguts of *Ligia italica* and *Ligidium hypnorum*. The area situated laterally to the primary filter is covered with long slender spines in both species and is surrounded by a posterior row of tooth-like spines in *Ligidium hypnorum* (Pl. 3B). In *Trachelipus illiyricus*, the lateralia are covered with sparsely arranged pectinate spines. Their ventral surface equipped with tooth-like spines is differentiated into the dorsal masticatory area which together with the ventral masticatory area adjacent to the primary filter, constitutes the masticatory apparatus (Pl. 3C, 8D). A paired ventral channels run laterally from the masticatory apparatus to the posterior part of the foregut (Pl. 1B).

The ultrastructure of the primary and secondary filters is similar in all four species, with slight difference in the position of filtering surfaces in relation to the longitudinal axis of the foregut. The foregut communicates with the midgut glands via the atrium situated ventrally below the secondary filter apparatus. In *Ligidium hypnorum* and *Ligia italica* the floor of the atrium consists of a small flap-like cuticular outgrowth of the inferomedianum and communicates with the midgut laterally (Pl. 4A, B). In *Porcellio scaber* and *Trachelipus illyricus* the atrium is large and sac-like, and communicates with the paired ducts of the midgut glands laterally and with the hindgut posteriorly (Pl. 4C).

3.2 *Foregut-hindgut junction*

A circular junction fold connects the posterior part of the foregut and the anterior chamber of the hindgut. This junction fold was described in *Porcellio scaber* as a posterior outgrowth of the atrium of midgut glands (Storch 1987) and in *Oniscus asellus* as a connection between the anterior openings of typhlosole channels and the

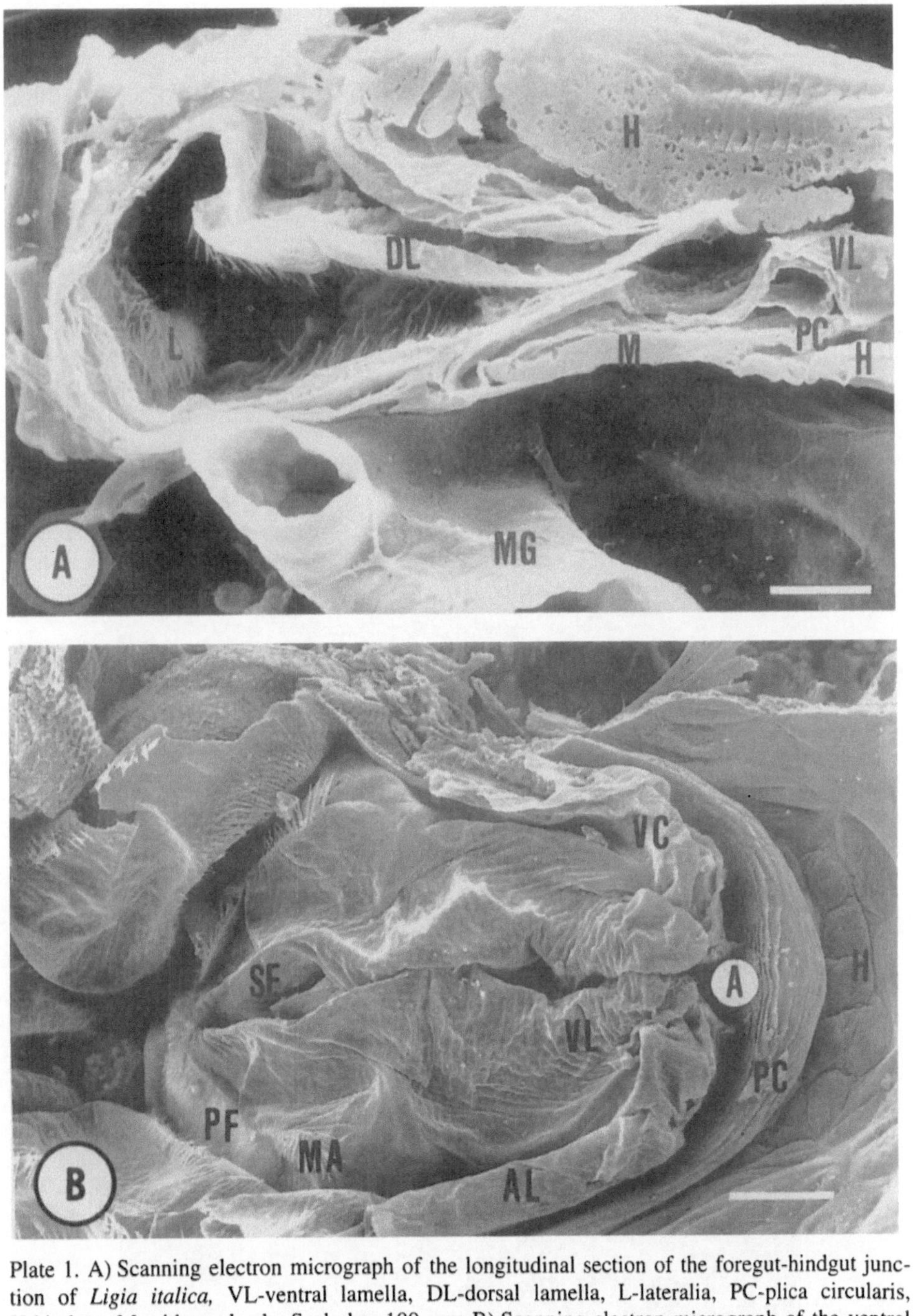

Plate 1. A) Scanning electron micrograph of the longitudinal section of the foregut-hindgut junction of *Ligia italica,* VL-ventral lamella, DL-dorsal lamella, L-lateralia, PC-plica circularis, H-hindgut, M-midgut glands, Scale bar 100 µm; B) Scanning electron micrograph of the ventral face of the foregut-hindgut junction of *Trachelipus illyricus*, VL-ventral lamella, AL-annular lamella, PF-primary filter, SF-secondary filter, MA-masticatory area, VC-ventral channel, A-atrium, PC-plica circularis, H-hindgut, Scale bar 100 µm.

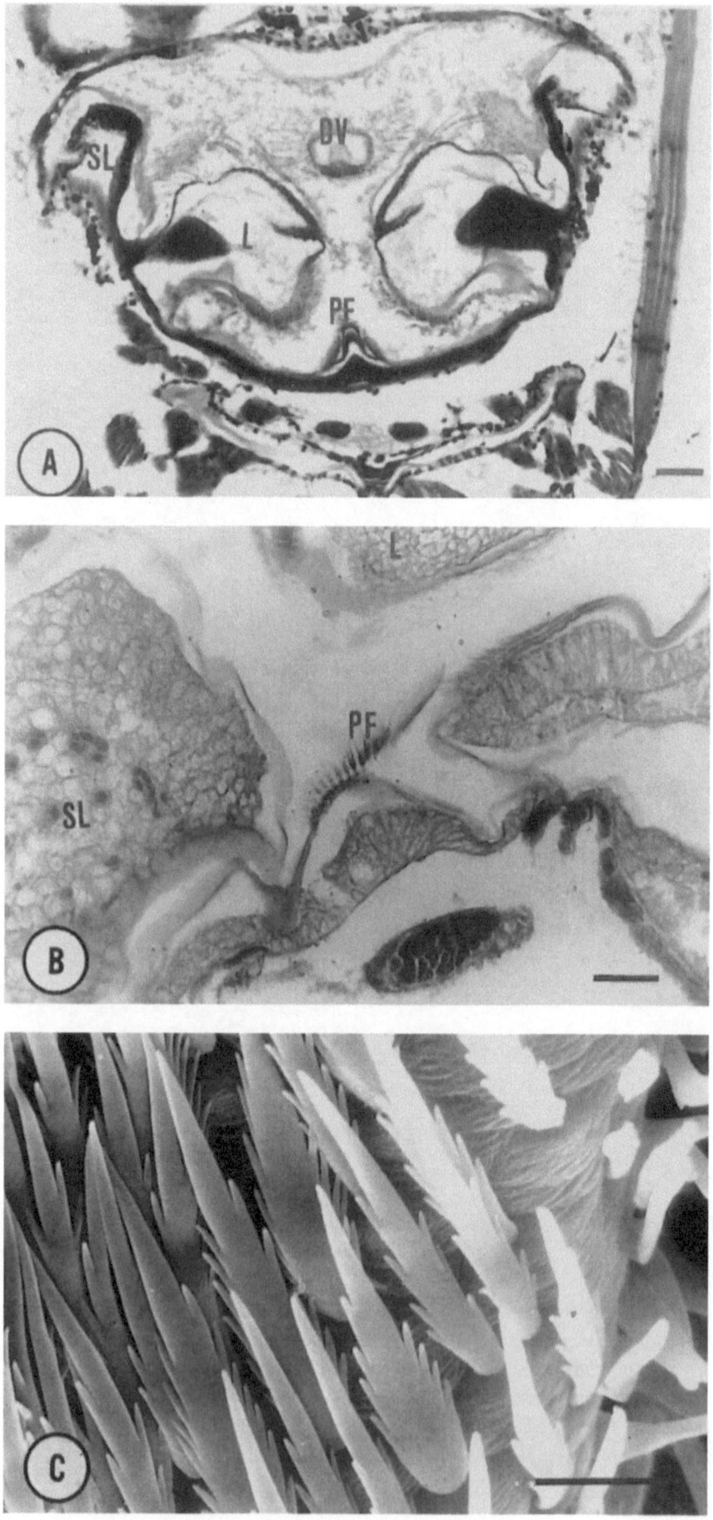
DV
SL
L
PF
A
L
PF
SL
B
C

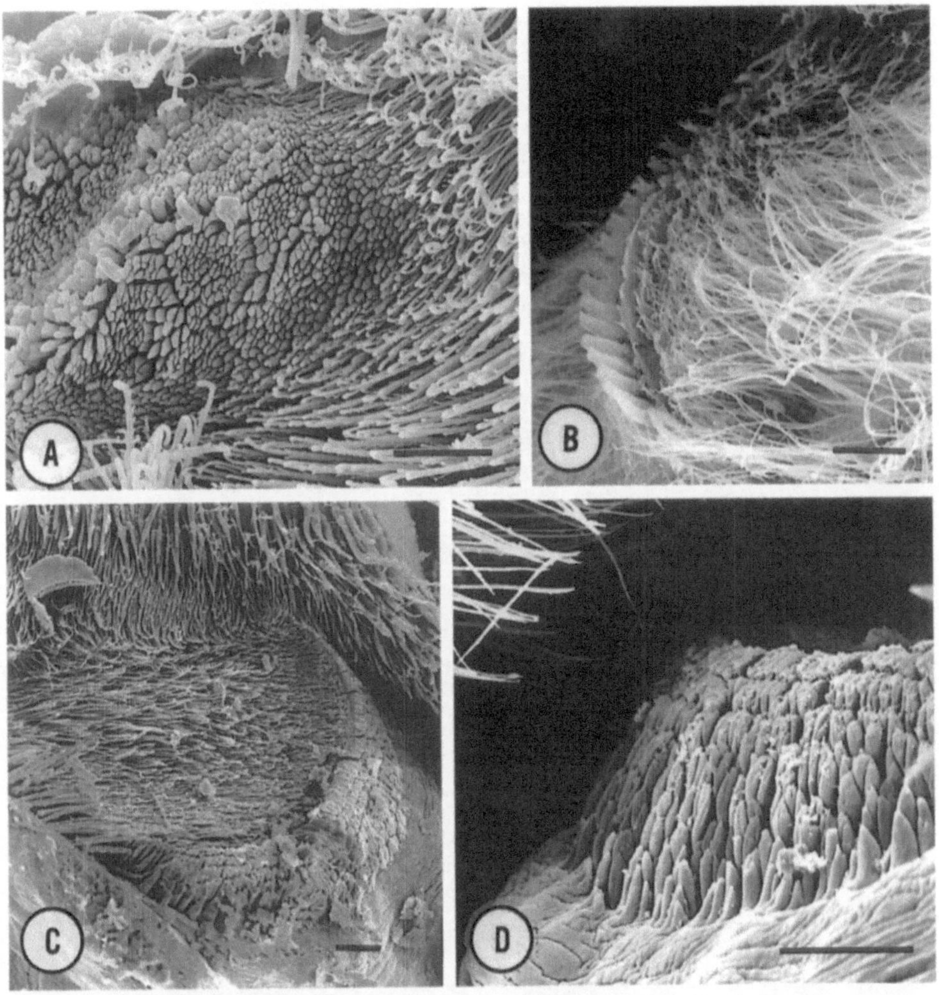

Plate 3. A) Scanning electron micrograph of the ventral face of lateralia in *Ligidium hypnorum;* B) Scanning electron micrograph of the surface situated laterally to the primary filter in Ligidium *hypnorum*, Scale bars 10 µm; C) Scanning electron micrograph of the ventral masticatory area in *Trachelipus illyricus* with several rows of compact teeth (d), Scale bars 10 µm.

Plate 2. A) Light micrograph of a cross section of the anterior foregut of *Ligia italica*, L-latyeralia, SL-superolateralia, PF-primary filter, DV-dorsal valve, Scale bar 10 µm; B) Light micrograph of longitudinal section of the foregut of *Ligidium hypnorum:* L-lateralia, SL-superolateralia, PF-primary filter, Scale bar 10 µm; C) Scanning electron micrograph of ventral spines of lateralia, Scale bar 5 µm.

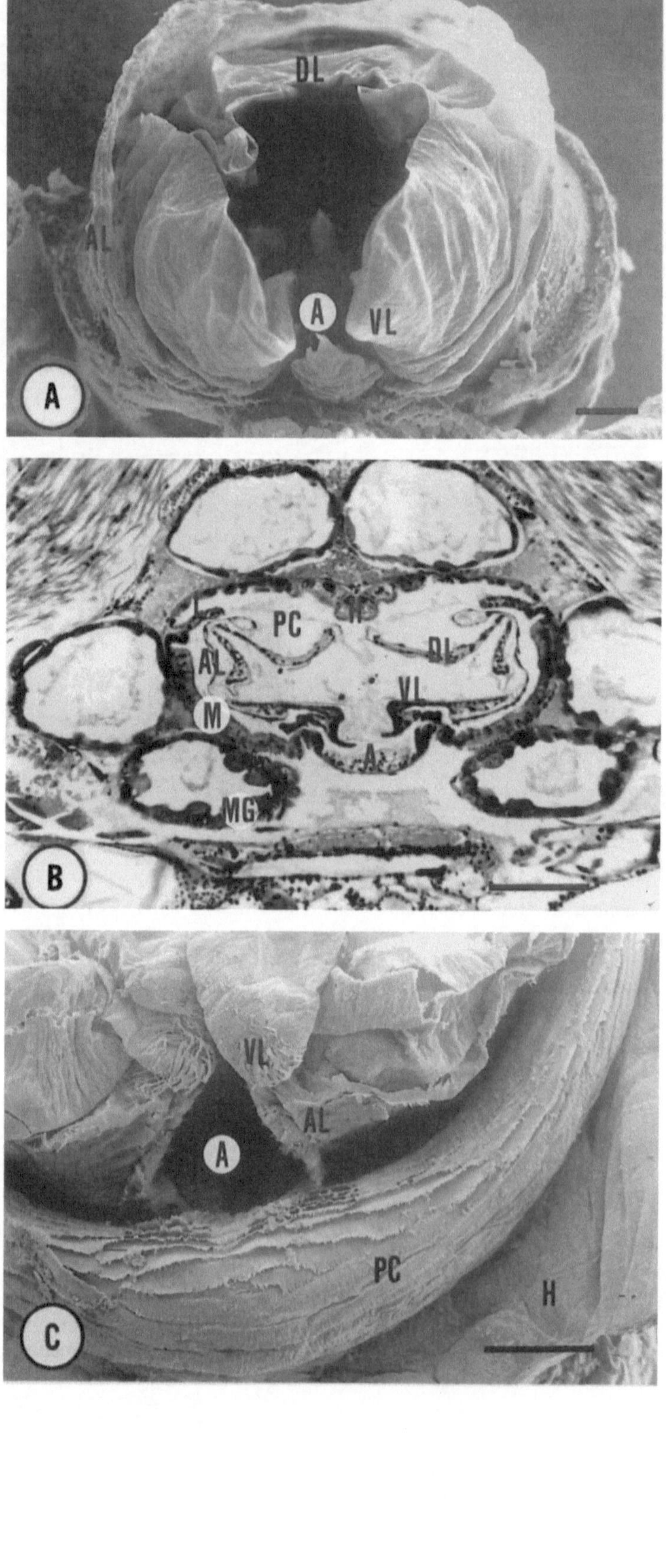

DL
AL
A
VL
A
L
PC
DL
AL
VL
M
A
MG
B
VL
AL
A
PC
H
C

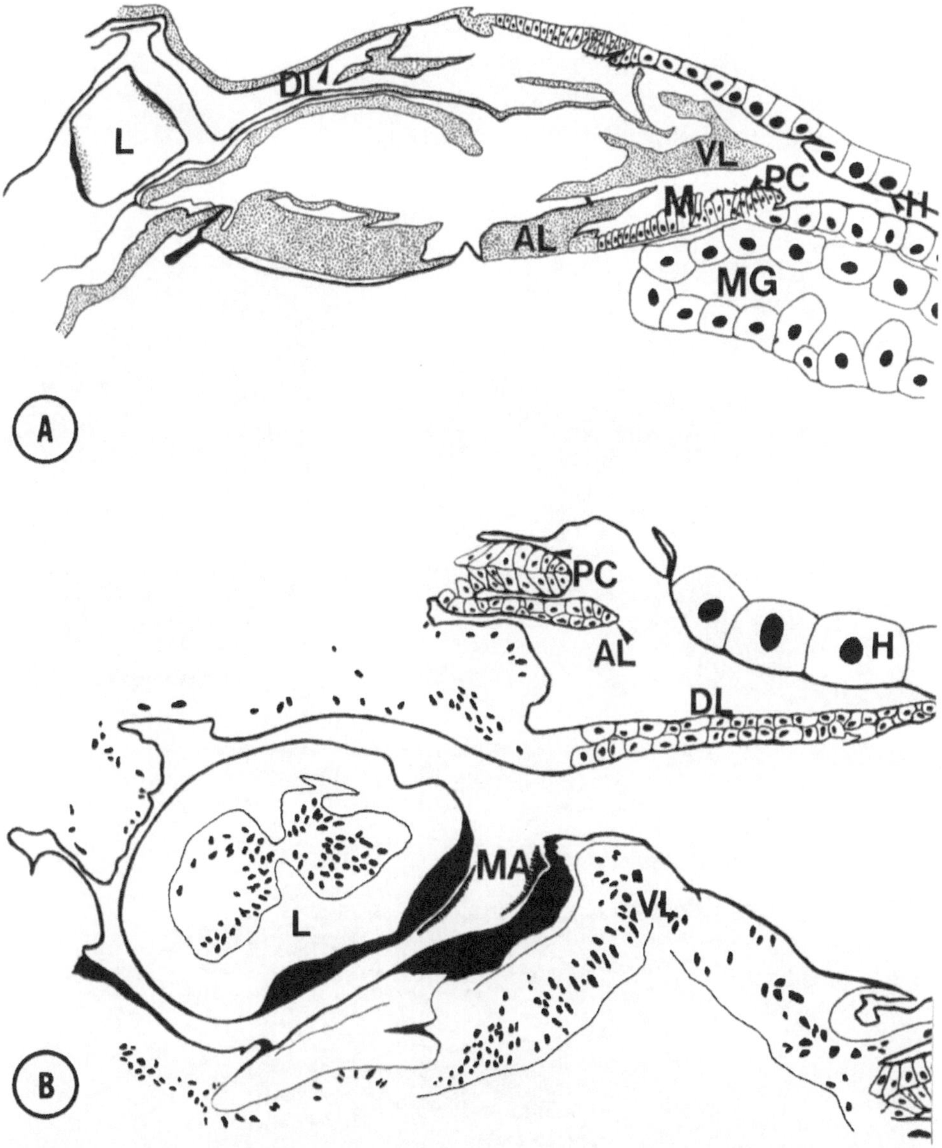

Plate 5. Schematic reconstruction of longitudinal sections of the foregut-hindgut junction of A) *Ligidium hypnorum* and B) *Oniscus asellus* (after Hames and Hopkin, 1989), L-, lateralia, MA-masticatory apparatus, DL-dorsal lamella, VL-ventral lamella, AL-annular lamella, PC-plica circularis, H-hindgut, M-midgut glands, Scale bars 100 μm.

Plate 4. A) Scanning electron micrograph of the posterior part of the foregut *of Ligidium hypnorum*, DL-dorsal lamella, VL-ventral lamella, AL-annular lamella, A-atrium, Scale bar 100μm; B) Light micrograph of cross section of the posterior foregut of *Ligia italica*, VL-ventral lamella, DL-dorsal lamella, AL-annular lamella, A-atrium, M-midgut, MG-midgut glands, PC-plica circularis, H-hindgut, Scale bar 50 μm; C) Scanning electron micrograph of the foregut-hindgut junction of *Trachelipus illyricus*, VL-ventral lamella, AL-annular lamella, A-atrium, PC-plica circularis, H-hindgut, Scale bar 5 μm.

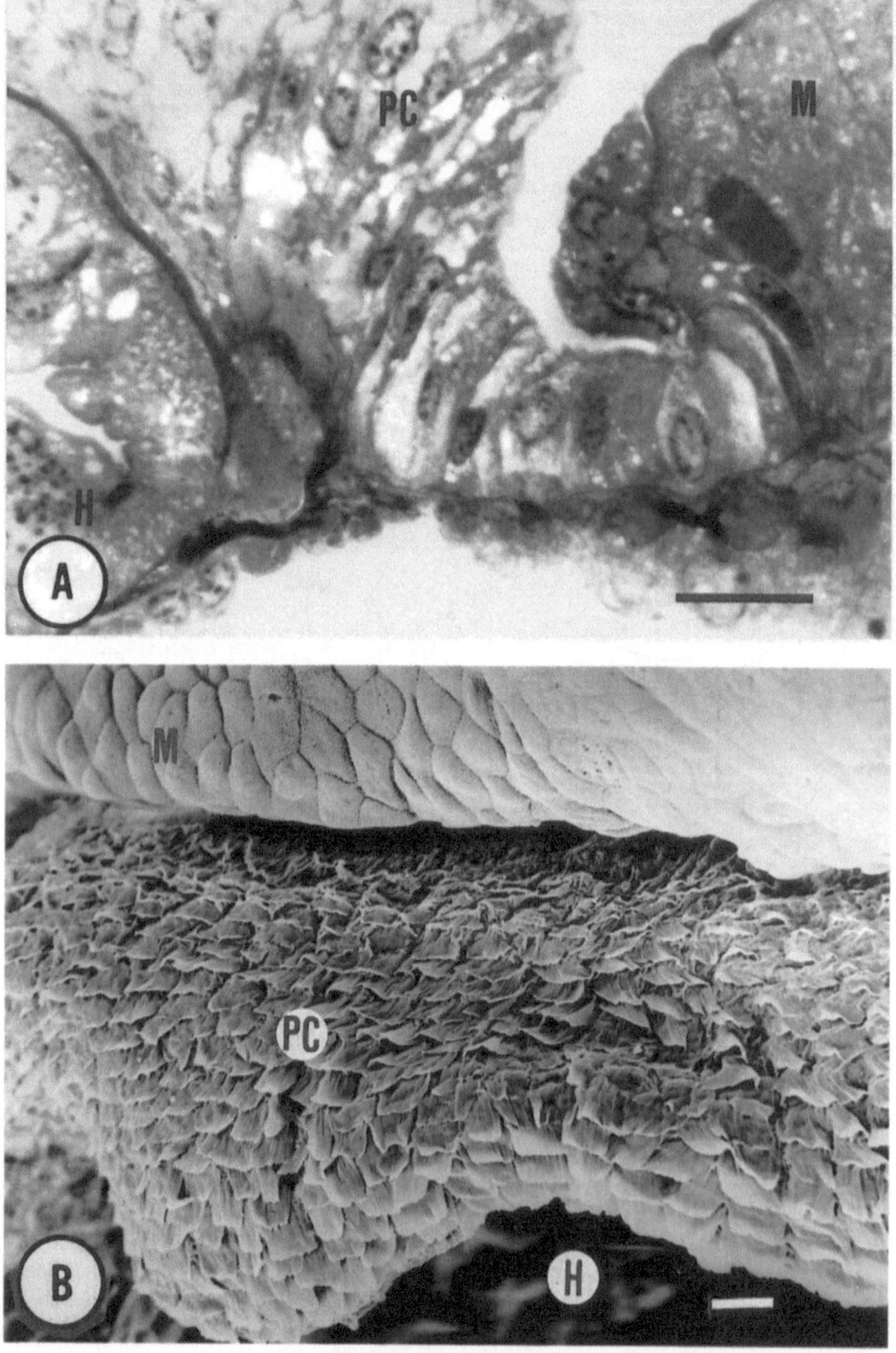

Plate 6. Midgut-hindgut junction of *Ligia italica*. A) Light micrograph of the midgut-hindgut junction, M-midgut, PC-plica circularis, H-hindgut, Scale bar 5 μm; B) Scanning electron micrograph of the midgut-hindgut junction, M-midgut, PC-plica circularis, H-hindgut, Scale bar 10 μm.

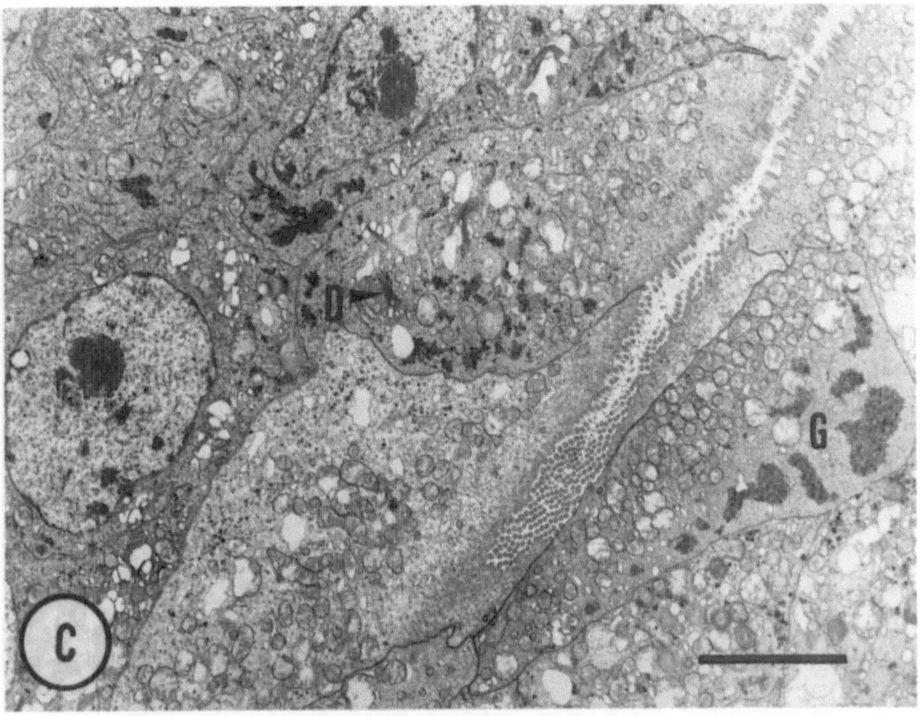

Plate 6. Continued. C) Transmission electron micrograph of midgut cells, D-dictyosomes, G-glycogen, M-mitochondria, Scale bar 5 μm.

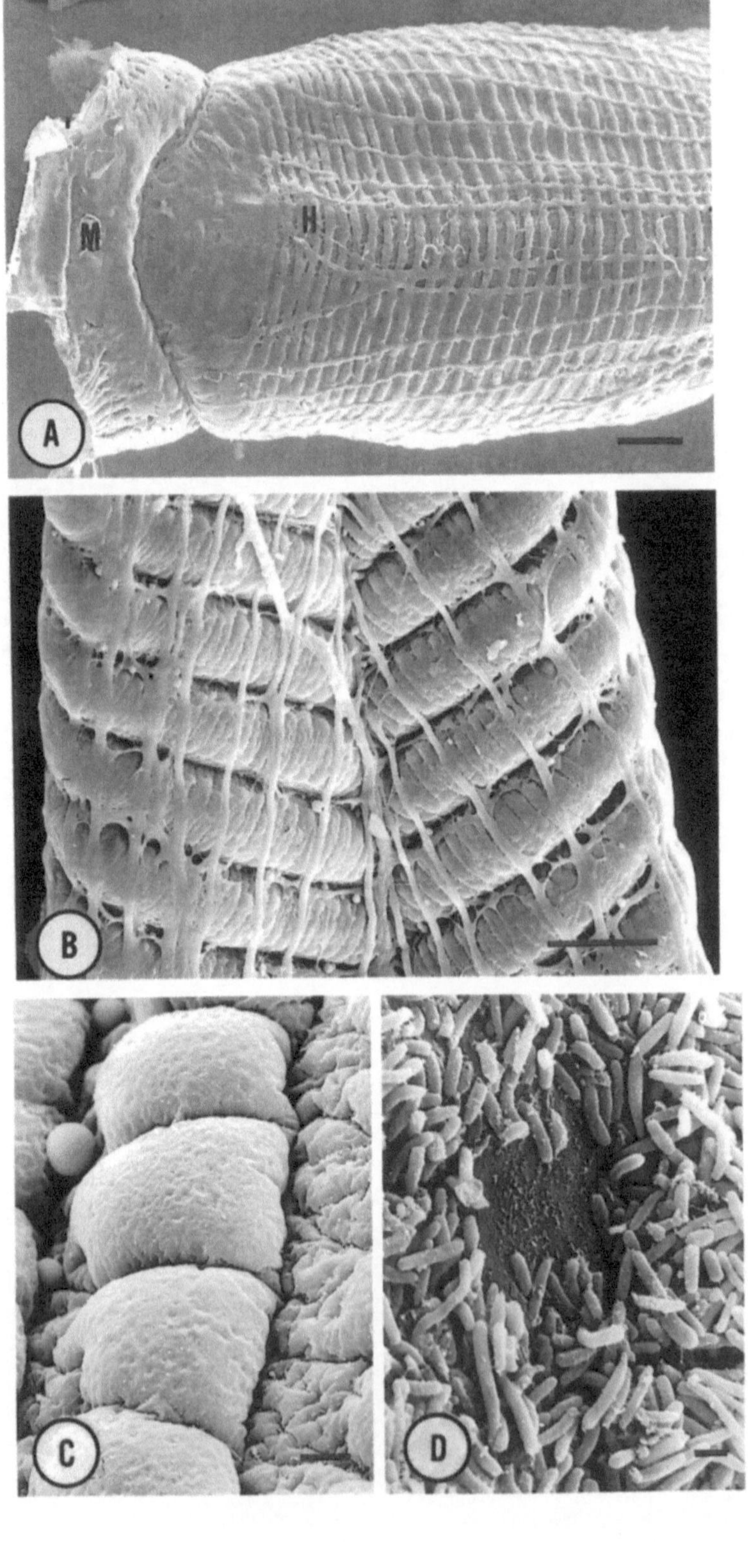
M
H
A
B
C
D

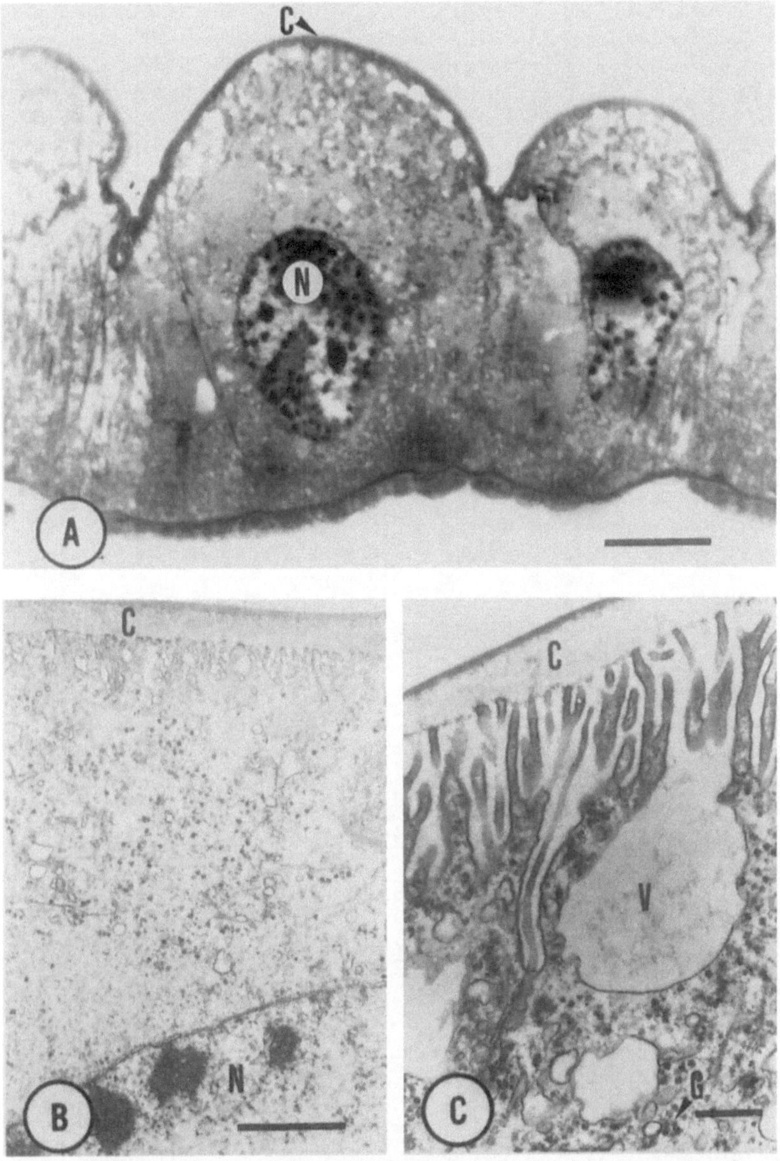

Plate 8. Hindgut of *Ligia italica.* A) Light micrograph of hindgut cells, C-cuticle, N-nucleus, V-vacuole, Scale bar 5 µm; B) Transmission electron micrograph of hindgut cell, N-nucleus, C-cuticle, Scale bar 2 µm; C) Transmission electron micrograph of hindgut cell, C-cuticle, V-vacuole, G-glycogen, Scale bar 1 µm.

Plate 7. A) Scanning electron micrograph of the anteries hindgut of *Ligidium hypnorum*, M-midgut, H-hindgut, Scale bar 100 µm; B) Scanning electron micrograph of the ventral face of the posterior hindgut of *Ligia italica*, Scale bar 100 µm; C) Scanning electron micrograph of the luminal surface of the hindgut of *Ligia italica*, Scale bar 10 µm; D) Scanning electron micrograph of rod-shaped bacteria on the luminal surface of the hindgut of *Ligidium hypnorum*, Scale bar 5 µm.

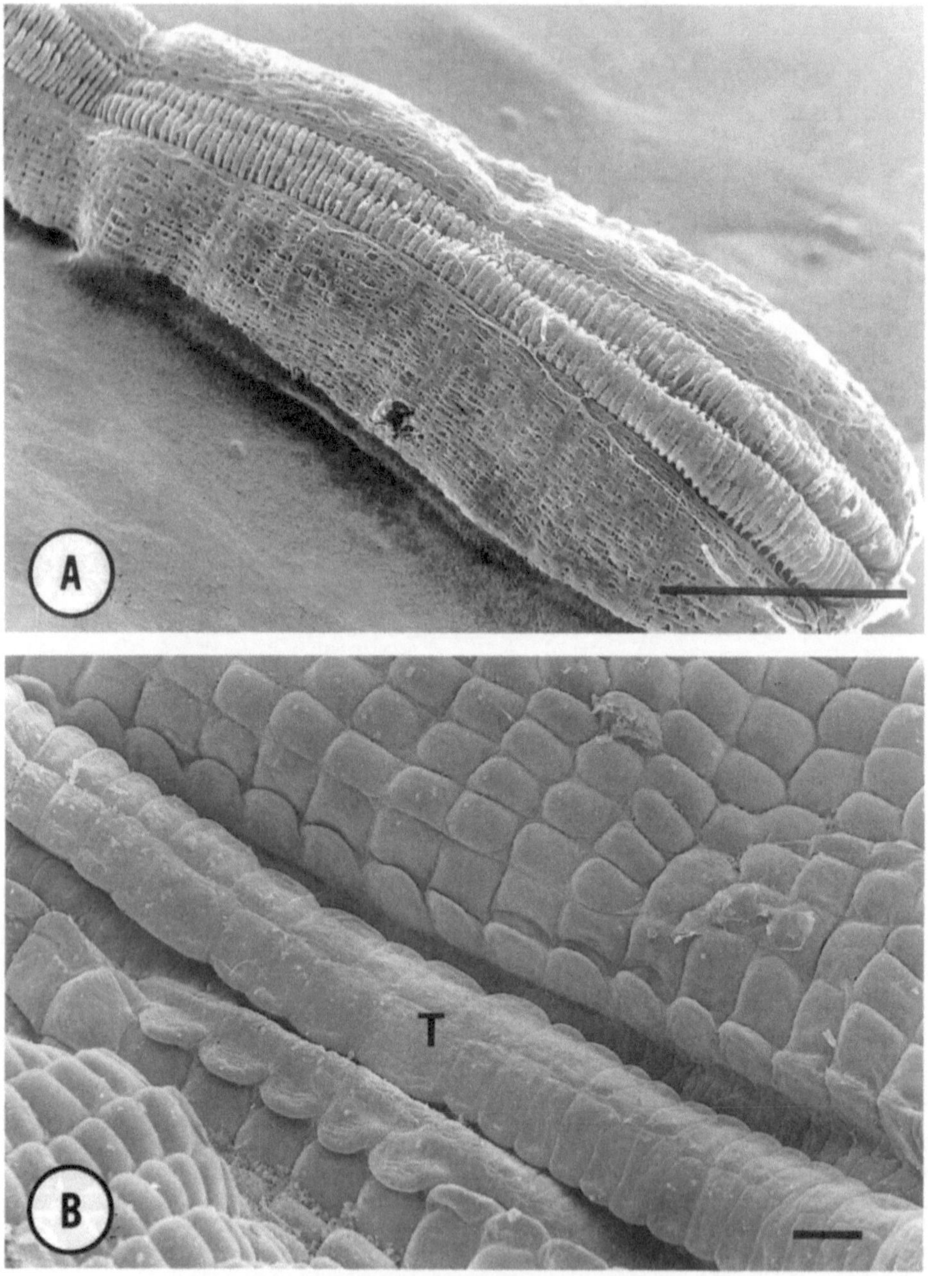

Plate 9. Anterior chamber of the hindgut of *Porcellio scaber*. A) Scanning electron micrograph of the muscular layer and typhlosolis, Scale bar 1mm; B) Scanning electron micrograph of the luminal surface of typhlosolis (T) with paired typhlosolis channels; Scale bar 100 μm.

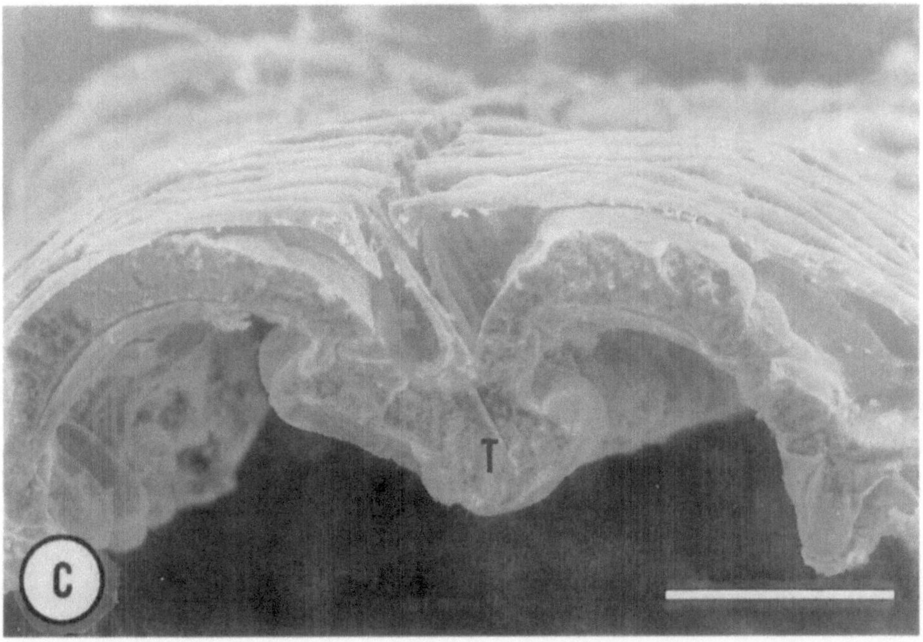

Plate 9. Continued. C) Scanning electron micrograph of the muscular layer of typhlosolis (T), Scale bar 100 μm.

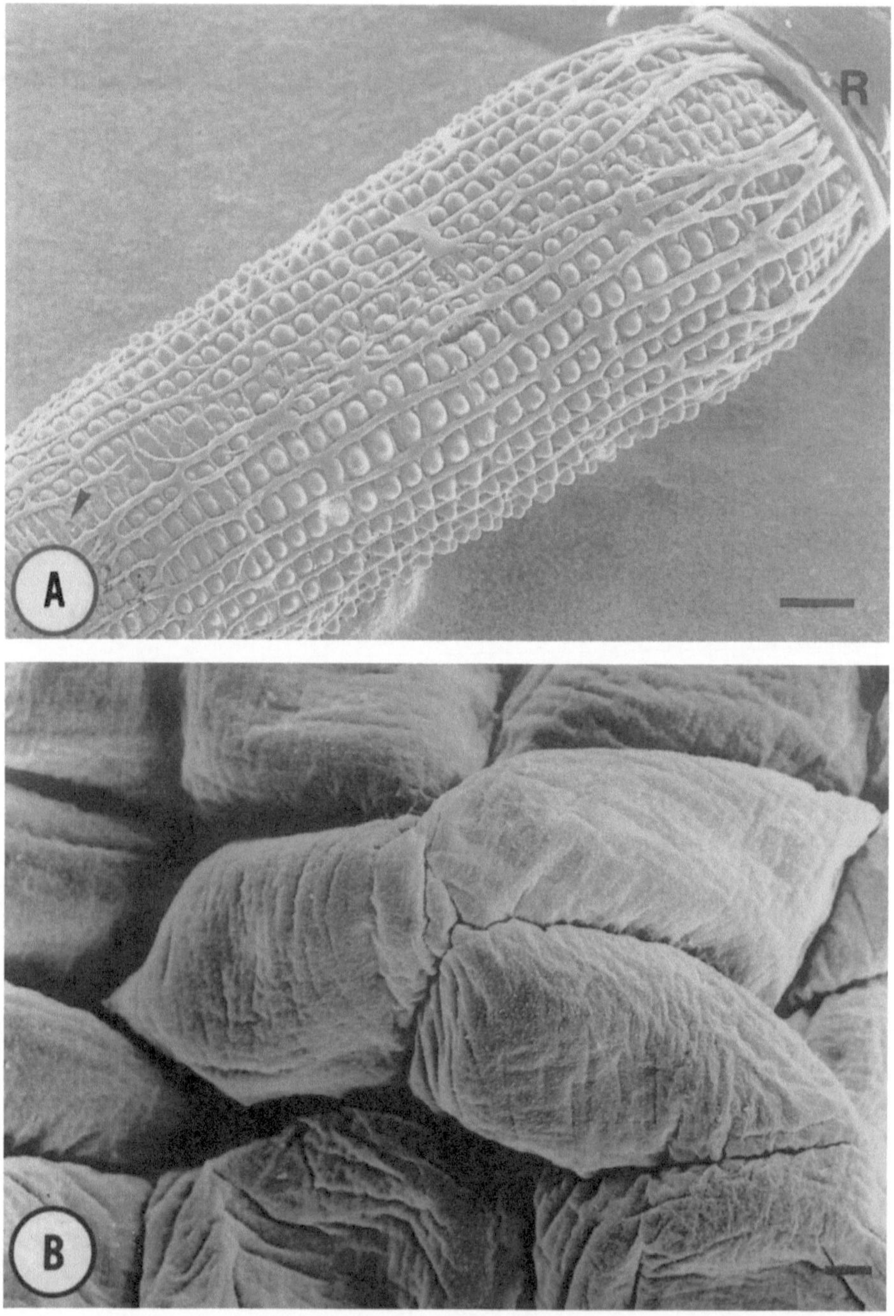

Plate 10. Posterior papillate region of the hindgut of *Porcellio scaber*. A) Scanning electron micrograph of the muscular layer, arrowhead marks the end of the typhlosolis, R-rectum, scale bar 100 µm; B) Scanning electron micrograph of the cuticular spines on the luminal surface of the papillate region, Scale bar 10 µm.

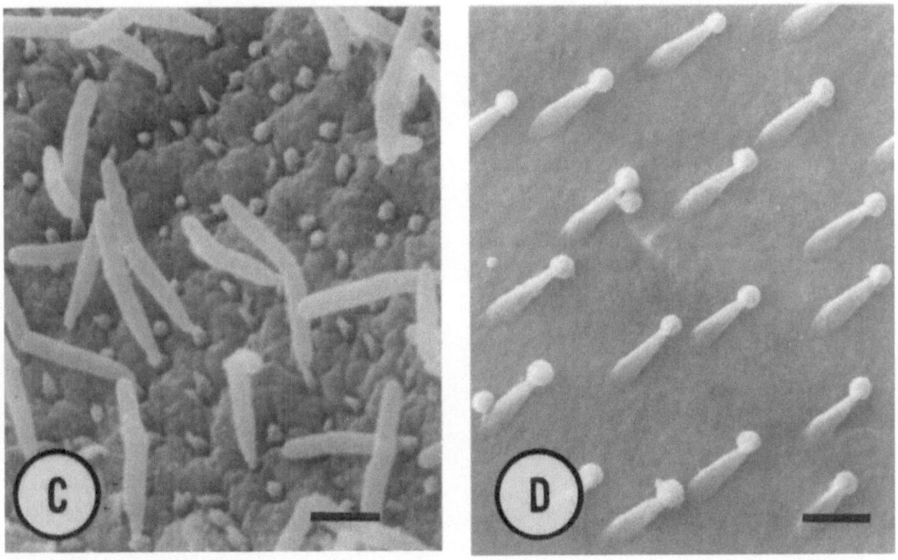

Plate 10. Continued. C) Bacteria attached to the spines, Scale bar 1 μm; D) Cuticular spines with spherical structures, Scale bar 1 μm.

atrium of midgut glands (Hames & Hopkin 1989). The microscopic anatomy of the foregut-hindgut junction, based on longitudinal sections of the alimentary canal of *Ligidium hypnorum* and *Oniscus asellus*, is presented in Plates 5. In the primitive species *Ligıa italica* and *Ligidium hypnorum*, the foregut is surrounded by a short endodermal midgut which is connected to the annular lamella of the foregut anteriorly and to the folded part of the hindgut posteriorly (Pl. 4B, 5A). A circular junctional fold separates the endodermal midgut from the anterior chamber of the hindgut (Pl. 6A). The epithelial cells of the fold are covered with cuticular comb-like structures arranged in parallel rows (Pl. 6B). The complex cuticular surface of the circular fold presumably acts as a filter which directs fluids to the hindgut and prevents coarse food particles from entering the midgut. The cells of the midgut, posessing a dense brush border contain numerous dictyosomes and secretory vesicles as well as large amounts of glycogen (Pl. 6C). The anatomy of the foregut-hindgut junction of *Trachelipus illyricus* is presented in (Pl. 4C). The cuticle of the foregut is continuous with the cuticle of the circular fold of the hindgut, and the endodermal part of the digestive system is restricted to two pairs of midgut glands.

3.3 *Hindgut*

The hindgut in all four species is subdivided into the anterior chamber, and the posterior region containing the rectum. The anterior part of the hindgut in *Ligia italica* and *Ligidium hypnorum* is encircled by a loose network of circular and longitudinal muscle fibers (Pl. 7A) while the posterior part has a thicker layer of circular muscles. The typhlosole is absent from the anterior hindgut of the two species. The densly packed cells of the ventral part of the posterior hindgut of *Ligia italica* are arranged in parallel rows which run obliquely to the longitudinal axis of the hindgut (Pl. 7B). The luminal surface is devoid of cuticular structures, but microorganisms attached to the hindgut cuticle were observed in both species (Pl. 7C, D). The homogeneous hindgut microflora consisted mainly of rod-shaped bacteria. The cells of the anterior chamber of the hindgut contain glycogen and have large nuclei with dense heterochromatin patches (Pl. 8A). The apical surface of the cells is invaginated and covered with a thin cuticle. Large vacuoles and mitochondria are concentrated in the apical cytoplasm of the cells (Pl. 8B, C).

The hindgut of terrestrial species is more complex. In *Porcellio scaber* the dorsal wall of the anterior chamber is invaginated and forms two typhlosolis channels which extend from the midgut gland duct junctions to the posterior papillate region (Pl. 9A, B). A thick layer of circular muscles runs from the lateral parts to the midline of the dorsal surface of the hindgut where they attach to the basal parts of the cells of the typhlosole (Pl. 9C). The papillate region of the hindgut is wrapped by a thick network of longitudinal and circular muscles. The basal parts of the cells bulge into the surrounding hemolymph when the muscles contract (Pl. 10A). The inner cuticular surface of the anterior hindgut is covered with short spines oriented posteriorly (Pl. 10B). The diverse cuticular spines of the papillate region are more numerous and often covered with bacteria some of which are attached to the spines (Pl. 10C). Small spherical structures are present at the tips of the spines in the posterior hindgut (Pl. 10D).

4 DISCUSSION

This comparative study of the digestive system of two amphibious and two strictly terrestrial isopods shows that significant differences exist in the structure of the ectodermal digestive tract, an important passage-way for nutrients as well as the site for water resorption and conservation. The anatomy of the foregut differs markedly between the species studied. The long flattened foregut of the two amphibious isopods and its luminal surface covered with long cuticular spines oriented posteriorly demonstrate that its main function is the transport of coarse food particles to the hindgut. Numerous cushion-like protrusions (ampullae) covered with spines press the food which then sifts through the two sets of filters.

The filtrate is conveyed to the atrium and to the midgut glands where it is absorbed (Štrus et al. 1985). The epithelial cells of the atrium are covered with a cuticle and communicate with the endodermal midgut laterally. The ultrastructure of the midgut cells with numerous dictyosomes and secretory vesicles suggests their secretory function. One of the possible functions of the midgut could be the secretion of digestive enzymes transported to the foregut, where food is partly degraded. The lamellae of the foregut which project deeply into the hindgut occlude the entrance to the midgut compartment and prevent the antiperistaltic flow of coarse particles from the hindgut.

The hindgut of amphibious species is a simple straight tube with a loose muscular network and without typhlosole. The luminal surface cover of small cuticular spines and the presence of abundant bacterial flora suggest that the final degradation of food is acomplished in the hindgut. The ultrastructure of the cells with deep invaginations of the apical membrane shows that absorption takes place in the anterior hindgut. The absorption of food in the anterior hindgut of *Mesidotea entomon* was demonstrated by autoradiographic analysis of cells of starved and fed animals (Hryniewiecka-Szyfter & Storch 1986). It is higly probable that bacteria present in the hindgut produce specific digestive enzymes, which degrade the food. Digestive juices are then absorbed in the anterior hindgut or returned to the atrium of the foregut. It was suggested by several authors that bacteria make the required nutrients more readily available and that they supplement food with trace elements, vitamins, and amino acids (Hassal & Rushton 1982, Carefoot 1984, Ulrich et al. 1991).

Amphibious isopods feed predominantly on algae, epiphytes and decaying organic material of a soft consistency. From the morphological data of the digestive tract we assume that the food is squeezed and filtered in the foregut. The filtrate is drained to the atrium and to the midgut glands, and the coarse particles are conveyed to the hindgut. It is probable that the secretory products from the midgut are transported to the hindgut where the food remains are enzymatically degraded and absorbed.

The results of this comparative study show that the digestive system of strictly terrestrial isopods is much more complex. The food, consisting of leaf litter, fresh plant material or soil particles, is triturated and filtered in the foregut. The filtrate is drained to the atrium and to the midgut glands. The coarse food particles which remain on the filters may be triturated in the masticatory apparatus and conveyed to the atrium via paired ventral channels. The results of previous studies of the isopod foregut show that the atrium of strictly terrestrial isopods communicates with all parts of the digestive system (Storch 1987, Hames & Hopkin 1989). Our study confirms these conclusions and suggests that the atrium is subdivided into an anterior and a posterior chamber that communicate with the midgut glands and hindgut respectively.

The foregut-hindgut junction is an important site for food channelling. The foregut is attached to the circular fold of the hindgut and a true midgut is absent in the terrestrial species included in the present study. The absence of a midgut was also demonstrated in *Oniscus asellus* and *Porcellio scaber* (Bettica et al. 1987). The hindgut of terrestrial isopods is a complex tube with several muscular layers. Its anterior part containing a prominent typhlosole plays an important role in the recycling of fluids (Hames & Hopkin 1989). The thick circular muscles attached to the basal part of the typhlosole cells extend the typhlosole and subdivide the upper part of the anterior hindgut into paired typhlosole channels which are completly isolated from the rest of the hindgut. The presence of several compartments in the anterior hindgut of terrestrial isopods results in partitioning of large food particles and digestive fluids and aids in absorption of nutrients. The posterior hindgut is encircled by a prominent muscular network. When the muscles contract, the basal parts of the hindgut cells bulge into the hemolymph. The ultrastructural analyses of the cells of the posterior hindgut of *Mesidotea entomon* and *Armadillidium vulgare* show that this part of the hindgut is involved mainly in ion and water transport (Hryniewiecka-Szyfter & Tyczewska 1979, Vernon et al. 1974).

The closed-type of water conducting system is an important adaptation of isopods to a terrestrial lifestyle (Hoese 1981). There are also some indications that posterior hindgut is involved in active water uptake (Verhoeff 1920, Palackal et al. 1984). The swollen basal parts of the cells of the papillate region and rectum support the assumption that osmoregulatory processes take place in this part of the hindgut. The role of the abundant bacterial flora in the hindgut has already been discussed, but it is worth mentioning that in strictly terrestrial isopods the bacteria are attached to the tips of the cuticular spines of the posterior hindgut. It is possible that the products of bacterial degradation are returned to the atrium via the typhlosole channels. Bacteria adherent to the hindgut of terrestrial isopods have been described by several authors (Drobne 1992, Griffiths & Wood 1985, Reyes & Tiedje 1976, Hassal & Rushton 1982) but information of their role in feeding strategies is still scant. We may conclude that the functional morphology of the digestive tracts of amphibious and strictly terrestrial isopods reflects the differences related to their lifestyles and feeding strategies. This study has shown that the presence of a complex ectodermal digestive tract and the reduction of its endodermally derived part in strictly terrestrial isopods, are important adaptations which have enabled isopods to invade such a variety of environments.

ACKNOWLEDGEMENTS

The authors would like to thank Dr. K. Drašlar and J. Rode for assistance with electron microscopy.

REFERENCES

Alikhan, M.A. 1969. The physiology of the *woodlouse Porcellio laevis* L. I.Studies on the gut epithelium cytology and its relation to maltase secretion. *Can. J. Zool.* 47:65-75.

Alikhan, M.A. 1972. Haemolymph and hepatopancreas copper in *Porcellio laevis* Latreille (Porcellionidae, Peracarida). *Comp. Biochem. Physiol.* 42A:823-832.

Bettica, A., R Witkus & G.M. Vernon 1987. Ultrastructure of the foregut-hindgut junction in *Porcellio scaber* Latreille. *J. Crust. Biol.* 7:619-623.

Bettica, A., M.T. Shay, G. Vernon & R. Witkus 1984. An ultrastructural study of cell differentiation and associated acid phosphatase activity in the hepatopancreas of *Porcellio scaber. Symp. Zool. Soc. Lond.* 53:199-215.

Carefoot, T.H. 1984. Nutrition and Growth of *Ligia pallassii. Symp. Zool. Soc. Lond.* 53:455-467.

Clifford, B. & E.R. Witkus 1971. The fine structure of the hepatopancreas of the woodlouse *Oniscus asellus. J. Morphol.* 135:335-350.

Drobne, D. 1993. Bacteria adherent to the hindgut of terrestrial isopods. In press.

Davies, S. & A. Forge 1987. Preparation of the mammalian organ of Corti for scanning electron microscopy. *J. Microscop* 147:89-101.

Donadey, C. 1973. Contribution a l'etude ultrastructurale et histophysiologique des caecums digestifs des crustaces isopodes. These Doc.d'Etat Sci.Nat. *C.N.R.S.* 7662:1-157.

Donadey, C. & G. Besse 1972. Etude histologique, ultrastructurale et expérimentale des caecums digestifs de *Porcellio dilatatus* et *Ligia oceanica* (Crustacea, Isopoda). *Tethys* 1:145-162.

Goodrich, A.L. 1939. The origin and fate of the entoderm elements in the embryogeny of *Porcellio laevis* Latr. and *Armadillidium nasatum* L.(Isopoda). *J. Morphol.*, 64:401-429.

Griffiths, B.S. & S. Wood 1985. Microorganisms associated with the hindgut of *Oniscus asellus* (Crustacea,Isopoda). *Pedobiologia* 28:377-381.

Hames, C.A.C. & S.P. Hopkin 1989. The structure and function of the digestive system of terrestrial isopods. *J. Zool. Lond.* 217:599-627.

Hames, C.A.C. & S.P. Hopkin 1991. A daily cycle of apocrine secretion by the B cells in the hepatopancreas of terrestrial isopods. *Can. J. Zool.* 69 1931.

Hassal, M. & J.B. Jennings 1975. Adaptive features of gut structure and digestive physiology in the terrestrial isopod *Philoscia muscorum* (Scopoli) 1763. *Biol. Bull.* 149:348-364.

Hassal, M. & S.P. Rushton 1985. The adaptive significance of coprophagous behaviour in the terrestrial isopod *Porcellio scaber. Pedobiologia* 28:169-175.

Hoese, B. 1981. Morphologie und Funktion des Wasserleitungssystems der terrestrischen Isopoden (Crustacea, Isopoda, Oniscoidea). *Zoomorphol.* 98:135-167.

Holdich, D.M. & K.R. Mayes 1975. A fine-structural re-examination of the so-called 'midgut' of the isopod *Porcellio. Crustaceana* 29:186-192.

Holdich, D.M. & N.A. Ratcliffe 1970. A light and electron microscope study of the hindgut of the herbivorous isopod, *Dynamene bidentata* (Crustacea:Peracarida). *Zeitschr. Zellforschung* 111:209-227.

Hopkin, S.P. 1986. The woodlouse *Porcellio scaber* as a biological indicator of zinc, cadmium, lead and copper pollution. *Environ. Poll.* (B)11:271-290.

Hryniewiecka-Szyfter, Z.& V. Storch 1986. The influence of starvation and different diets on the hindgut of Isopoda (*Mesidotea entomon, Oniscus asellus, Porcellio scaber*). *Protoplasma* 134:53-59.

Hryniewiecka-Szyfter, Z. & J. Tyczewska 1979. Fine structure and localization of alkaline phosphatase in the hindgut of *Mesidotea entomon* L. (Isopoda,Crustacea). *Bull. Soc. Amis Sci. Lett. Poznan, S r.D,* 19:57-64.

Hubschman, J.H. 1962. A simplified Azan process well suited for crustacean tissue. *Stain technol.* 379-380.

Jones, D A 1968. The functional morphology of te digestive system in the carnivorous intertidal isopod *Eurydice. J.Zool.* 156(3):363-377.

Jones, D.A., P.C. Babbage & D.E. King 1969. Studies on the digestion and the fine structure of digestive caeca in *Eurydice pulchra* (Crustacea, Isopoda). *Mar.Biol.* 2:311-320.

Lane, L.R. 1988. Thee digestive system of *Porcellio scaber* Latreille, 1804 (Isopoda, Oniscoidea): Histology and histochemistry. *Crustaceana* 55:113-128.

Ličar, P. 1975. Prispevek k poznavanju zgradbe in funkcije prebavila pri rasah *Asellus aquaticus* (Isopoda,Asellota). *Razprave SAZU, Classis IV:Historia naturalis* 18:157-203.

Palackal, T., L. Faso, J.L. Zung, G. Vernon & R. Witkus 1984. The ultrastructure of the hindgut epithelium of terrestrial isopods and its role in osmoregulation. *Symp. Zool. Soc. Lond.* 53:185-198.

Prosi, F. & R. Dallinger 1988. Heavy metals in the terrestrial isopod *Porcellio scaber* Latreille. I.Histochemical and ultrastructural characterization of metal-containing lysosomes. *Cell. Biol. Toxicol.* 4/1:81-96.

Prosi, F., V. Storch & H.H. Janssen 1983. Small cells in the midgut glands of terrestrial Isopoda: Sites of heavy metal accumulation. *Zoomorphology* 102:53-64.

Reyes, V.G. & J.M. Tiedje 1976. Ecology of the gut microbiota of *Tracheoniscus rathkei* (Crustacea, Isopoda). *Pedobiologia* 16:67-74.

Spurr, A.R. 1969. A low viscosity embedding medium for electron microscopy. *J. Ultrastruct. Res.* 26:31-43.

Storch, V. 1982. Der Einfluss der Ernahrung auf die Ultrastruktur der grossen Zellen in den Mitteldarmdrüsen terrestrischer Isopoda. *Zoomorphology* 100:131-142.

Storch, V. 1984. The influence of nutritional stress on the ultrastructure of the hepatopancreas of terrestrial isopods. *Symp. Zool. Soc. Lond.* 53:167-184.

Storch, V. 1987. Microscopic anatomy and ultrastructure of the stomach of *Porcellio scaber* (Crustacea,Isopoda) *Zoomorphology* 106:301-311.

Storch, V. & J. Štrus 1989. Microscopic anatomy and ultrastructure of the alimentary canal in terrestrial isopods. *Monitore Zool. Ital. (N.S.) Monogr.* 4:105-126.

Štrus, J. 1987. The effects of starvation on the structure and function of the hepatopancreas in the isopod *Ligia italica*. *Inv. Pesq.* 51 (supl.1) 505-514.

Štrus J. & K. Drašlar 1988. Ultrastructural evidence of midgut cells in the isopod *Ligia italica* (Isopoda, Crustacea). *Inst. Phys. Conf. Ser.* 93:149-150.

Štrus, J., P. Burkhardt & V. Storch 1985. The ultrastructure of the midgut glands in *Ligia italica* (Isopoda) under different nutritional conditions. *Helgoländer Wiss. Meeresunters.* 39:367-374.

Tuzet, O., J.F. Manier & R. Ormieres 1959. Recherches sur l'appareil digestif de quelques Isopodes (anatomie, histologie, cytologie). *Bull. Soc. Zool. France* 84:505-531.

Ulrich, B., V. Storch & H. Schairer 1991. Bacteria on the food, in the intestine and on the faeces of the woodlouse Oniscus asellus (Isopoda, Crustacea). *Pedobiologia* 35:41-51.

Vandel, A. 1960. Isopodes terrestres. *Faune de France* 64:1-416.

Verhoeff, K.W. 1920. Über die Atmung der Landasseln, zugleich ein Beitrag zur Kentniss der Entstehung der Landtiere. *Z. Wiss. Zool.* 118:365-447.

Vernon, G.M., L. Herold & E.R. Witkus 1974. Fine structure of the digestive tract epithelium in the terrestrial isopod, *Armadillidium vulgare*. *J. Morph.* 144:337-360.

Wägele, J.W. 1989. Evolution und phylogenetisches System der Isopoda. *Zoologica* 140:1-262.

Wägele, J.W., U. Welsch & W. Müller 1981. Fine structure and unction of the digestive tract of *Cyathura carinata* (Krøyer) (Crustacea,Isopoda). *Zoomorphology* 98:69-88.

Warburg, M.R. & M. Rosenberg 1989. Ultracytochemical identification of Na+K- ATP-ase activity in the isopodan hindgut epithelium. *J. Crust. Biol.* 9:525-528.

Wieser, W., G. Busch & L. Büchel 1976. Isopods as indicators of the copper content of soil and litter. *Oecologia* 23:107-114.

Water vapour absorption and ammonia volatilization: Adaptations for terrestriality in isopods

Jonathan C. Wright
Department of Mathematical & Natural Sciences, Northern State University, Aberdeen, South Dakota, USA

Michael J. O'Donnell
Department of Biology, McMaster University, Hamilton, Ontario, Canada

ABSTRACT

A capacity for active water vapour absorption (WVA), a phenomenon hitherto described only for acarines and insects, has recently been demonstrated in terrestrial isopods (Isopoda, Oniscidea). Rapid vapour uptake in humid diurnal retreats allows many species to replenish substantial water losses sustained during nocturnal foraging. In theory, this should permit exploitation of relatively low foraging humidities. The uptake rates in saturated air, when expressed as g H_2O h^{-1} or as g H_2O h^{-1} per unit absorbing surface area, are the highest documented for known vapour absorbers. Absorption takes place in the pleon, and is accomplished by two complimentary mechanisms. Secretion of strongly hyperosmotic NaCl (up to 8.2 Osm kg^{-1}) into the pleoventral cavity generates a colligative lowering of vapour pressure, permitting vapour absorption above threshold humidities of 86.6% to 92.7% RH. Uptake rates can be increased and thresholds lowered by an additional facultative process involving compression of air beneath the imbricate pleopodal exopods. This serves to elevate the humidity of the enclosed volume of air, and represents the first demonstration of pressure cycling as a vapour absorption mechanism. Nitrogenous excretion, like WVA, plays an important role in the water economy of terrestrial isopods. Unlike the majority of insects and arachnids, they have retained ammonotelic excretion, characteristic of their marine ancestors. They thus constitute an exception to the broadly accepted dogma that ammonotely is restricted to hygric or aquatic groups. Although elimination of ammonia in gaseous form represents a means of minimizing concomitant water loss, the need to retain permeable regions of the integument for NH_3 diffusion inevitably compromises water conservation. The site and mechanism of ammonia volatilization have previously been the subject of controversy. Recent work has established a clear association between periods of ammonia volatilization and the occurrence of highly elevated concentrations of ammonia in both the haemolymph and the external pleon fluid which bathes the respiratory endopods. Volatilization appears to be driven by the large concentration gradients between pleon fluid NH_3 and the atmosphere; the pNH_3 is not augmented by alkalinization of pleon fluid. Bouts of ammonia volatilization also show a clear temporal association with metachronal beating of the pleopods. The substantial bursts of haemolymph ammonia coincident with volatilization require the rapid mobilization of ammonia from a precursor. Current studies indicate a primary role for arginine, glycine, glutamine and glutamate as precursors. Several lines of evidence support a possible physiological coupling between ammonia volatilization and WVA. It is probable that vapour uptake is primarily diurnal, as has been clearly documented for ammonia volatilization. Volatilization has not been observed in humidities below the thresholds for WVA and, like WVA, is invariably associated with pleopodal ventilation. Coupling of the two processes, whether obligate or facultative, would provide a unique example of ammonotely actually being associated with a net gain of water. Possible evolutionary scenarios favouring retention of ammonotely along with the origins of water vapour absorption in isopods are discussed.

1 INTRODUCTION

Terrestrial isopods exploit both behavioural and physiological adaptations to enhance survival in potentially desiccating environments. The prevailing dogma, emerging largely from work in the 50's and 60's, is that behavioural adaptations predominate. To quote from Edney (1968) for example: 'Poorly equipped as most isopods are to resist the effects of dry air and high temperatures, sensory and behavioural mechanisms that permit them to avoid such conditions to a greater or lesser extent assume added significance.' Similarly, Cloudsley-Thompson (1956) states: 'There are two ways in which small animals can (avoid desiccation) on dry land. One is to (remain) in a humid environment as a result of physiological orienting mechanisms; the other to acquire a relatively impermeable integument and the physiological and morphological specializations that allow respiration, excretion and so on to take place without excessive loss of water. The writer (asserts) that the terrestrial Isopoda represent a group that has exploited the first of these methods.'

Recent studies, including our own, challenge such views, and have provided new evidence for sophisticated physiological mechanisms for withstanding water stress. Novel processes for water uptake and water conservation during nitrogenous excretion appear particularly significant, and constitute the theme of this paper.

1.1 *Water vapour absorption (WVA)*

The possibility of active water vapour absorption (WVA) in terrestrial isopods (Isopoda, Oniscidea) was first investigated by Spencer & Edney (1954). On the basis of intermittent weighings of animals in a range of humidities they were unable to demonstrate net increases in mass when ambient water activity (a_w = RH/100) was below that of the hemolymph. [The activity of water vapour in air is equal to RH/100, where RH is the relative humidity (%). An activity of 1, therefore, corresponds to saturation. The activity of water in aqueous solutions is approximately equal to the mole fraction of water present (55.5/55.5 + x), where x os the osmotic concentration of solutes in Osm kg^{-1}, and 55.5 the molal concentration of water). At a given activity, the liquid and vapour phases of water are in equilibrium. For example, an animal whose hemolymph osmolality is 1 Osm kg^{-1} has a hemolymph activity of 0.982 (=55.5/55.5 + 1), and will be in passive equilibrium with an ambient humidity of 98.2% RH. At lower humidities, the animal will loose water.] However, conflicting data were presented by Den Boer (1961) who showed a capacity for weight-increase in humidities down to ca. 91% in *Porcellio scaber*. Further studies by Coenen-Stass (1981, 1989) demonstrated a similar capacity for weight gain in the desert species *Hemilepistus afghanicus* in humidities above 93% RH. He proposed a process of vapour absorption but did not invoke an active mechanism. Subsequently, Wright & Machin (1990), using continuous weighing of isopods in controlled temperatures and humidities, provided unambiguous evidence for active uptake in three common temperate species: *Oniscus asellus, Porcellio scaber* and *Armadillidium vulgare*. WVA has subsequently been confirmed in 12 species of the oniscidean section Crinocheta, containing the familiar woodlice or sowbugs, and for one species of the section Diplocheta, *Ligia oceanica* (Wright & Machin 1993a).

In all species studied, maximum rates of vapour uptake increase approximately linearly as a function of ambient water activity above a minimum threshold (Fig. 1). It is also evident from Figure 1 that animals can regulate uptake rates to any level between zero and the maximum for a given ambient activity. Passive losses in non-absorbing animals show an inverse linear relationship to ambient water activity (Fig. 1). Because isopods sustain relatively high transpiratory water losses, the threshold activity for WVA is substantially lower than the critical equilibrium activity (CEA) at which gains and losses of water are balanced (Fig. 1). A detailed analysis of uptake kinetics is presented in Wright & Machin (1993b).

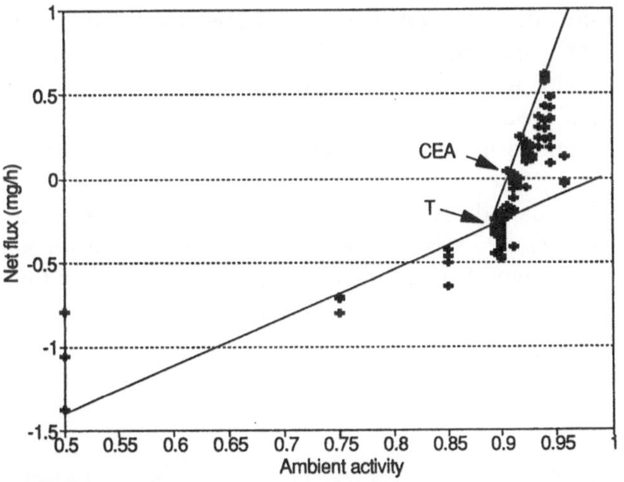

Figure 1. Plot showing the relationship between net water flux (mg h^{-1}) and ambient water activity (RH/100) for a vapour-absorbing oniscidean. Each data point represents a water flux measured over a 1-2 hour period. Passive losses (negative fluxes) are directly proportional to ambient activity below the haemolymph activity (ca. 0.988). The lower line (passive losses) was fitted by regression, omitting assumed instances of absorption and periods when loss fluxes were elevated by locomotory activity. The upper line was fitted by eye to the maximum net fluxes (absorption) measured at each ambient activity. Above a threshold activity of ca. 0.89, active WVA is possible and uptake rates can exceed passive losses above the critical equilibrium activity (CEA; ca. 0.91). The upper line was fitted by eye to maximum net flux measured at each ambient activity. Maximum uptake fluxes show an approximately linear increase with ambient activity, indicating a non-saturating absorption mechanism, although sub-maximal uptake fluxes are also possible. From Wright & Machin (1992a).

Experiments in which regions of the body were sealed by application of beeswax indicated that vapour uptake occurs across a specific region of the body surface, the ventral pleon. During vapour uptake, animals invariably display regular metachronal beating of the pleopods. This behaviour is referred to as ventilation, and presumably serves primarily to circulate humid air across the absorbing surfaces.

The 5 pairs of pleopods in oniscideans form a series of imbricating plates closely apposed to the ventral sternites. The exopods, primarily for gas exchange, conceal the endopods (as copulatory organs in pleopods 1 and 2), which thus lie within a chamber, usually referred to as the pleoventral cavity or 'pleoventralraum' (PV). Fluid secreted within the PV serves to maintain a permanent, thin film over the endopods. In many species the endopods possess specializations for respiratory exchange, thereby supplementing the role of the exopods (Verhoeff 1917, 1920; Unwin 1932). Fluid within the PV can be sampled with micropipettes (Fig. 2) and is isosmotic to the haemolymph in non-absorbing animals (Wright & O'Donnell 1992). Over the course of a few minutes prior to the onset of ventilation and vapour uptake, the isosmotic fluid is replaced with a copious, strongly hyperosmotic 'uptake fluid' (*ibid*). Freezing point depression measurements of nanoliter samples have revealed initial fluid osmolalities as high as 8.2 Osm kg^{-1}. Following the onset of WVA, the fluid osmolality declines to values representing water activities only slightly below ambient (Fig. 3). These observations are consistent with dilution of the uptake fluid by absorbed water. Absorption depends upon the colligative lowering of vapour pressure generated by elevated solute concentrations in the uptake fluid. The presence of hyperosmotic pleon fluids during WVA has been independently verified by melting point determinations using sections cut from fast-frozen, absorbing animals (Wright & Machin 1990). A model analysing the components of the colligative uptake mechanism in the context of observed uptake kinetics is presented in Wright & Machin (1993b). Pleon fluids, unlike haemolymph, show no clotting reaction and lack haemocytes.

The uptake fluid osmolytes consist primarily of Na^+ and Cl^-. Activities of Na^+ and Cl^-, determined in nanoliter samples by ion-selective microelectrodes, each account for 40-50% of total osmolality (Wright & O'Donnell 1992). Contributions of these and other electrolytes to WVA mechanisms in other arthropods have been reviewed

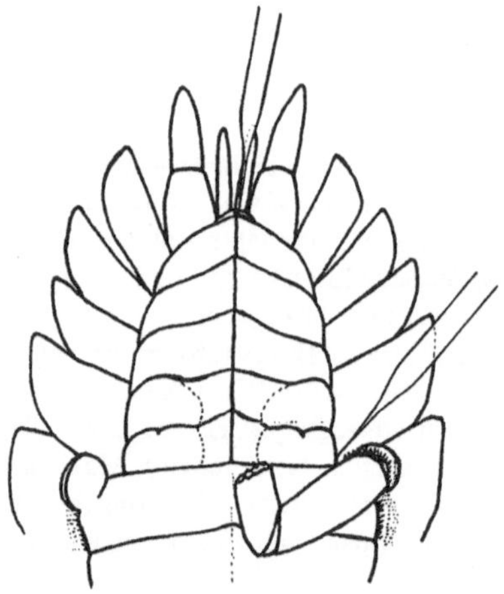

Figure 2. Semi-schematic diagram of the ventral pleon of *P. scaber* illustrating the 5 pairs of imbricate pleopodal exopods which overlie the fluid-bathed pleoventral cavity (PV) and endopods. The micropipettes illustrate favourable locations for sampling uptake fluid.

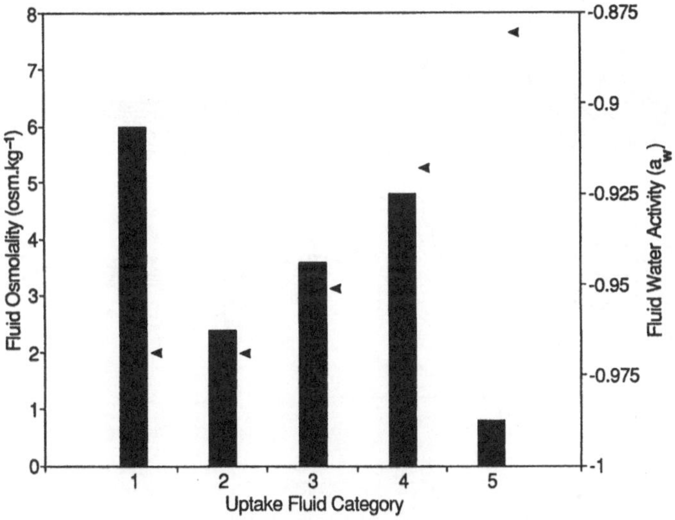

Figure 3. Schematic plot, based on data in Wright & O'Donnell (1992), showing the relationship between uptake fluid osmolality/activity and ambient activity (arrows) during WVA. Category 1 shows strongly hyperosmotic 'pre-ventilatory' fluid, with an activity much lower than ambient. Following the onset of ventilation, fluid osmolality falls to give activities only slightly below ambient (2 and 3). Category 4 illustrates an example where ambient activity is actually lower than that of the uptake fluid, and increases of water vapour activity within the PV by means of pressure cycling (see text) are inferred. In ambient activities below the uptake threshold, uptake fluid is resorbed, leaving a scant fluid, isosmotic with the haemolymph (5).

by O'Donnell & Machin (1988). Indirect evidence suggests that uptake fluid is probably secreted by the endopods. They possess classic transporting epithelia with apical microvilli and deep basal infoldings associated with abundant mitochondria (Kummel 1981, 1984). Immuno-labelling techniques also reveal very high ATPase activities in *Porcellio scaber* and 6 other terrestrial oniscideans (Wright, Holliday & O'Donnell 1994), consistent with a capacity to transport ions against large electrochemical gradients. No comparable transporting epithelia are found on other external surfaces. Whether the endopods are also involved in fluid resorption, or whether this involves a rectal conduit, remains unclear (see Wright & Machin 1993b).

1.2 *Pressure cycling*

A facultative mechanism for augmenting uptake rates involves the compression of enclosed air within the PV by the ventilating pleopods. Such a mechanism was suggested by Wright & O'Donnell (1992) to account for the occasional presence in absorbing animals of pleon fluid osmolalities which, although strongly hyperosmotic to the haemolymph, were insufficient to generate a water activity below ambient (see Fig. 4). Compression of the pleoventral air space would generate a corresponding increase in the water vapour density (mg cm^{-3}) and hence the activity. Water activity could thereby be elevated above that of the uptake fluid, allowing animals to maintain colligative absorption (Fig. 4). The process is referred to as 'pressure cycling', following Corbet (1988). The proposed mechanism in oniscideans is supported by

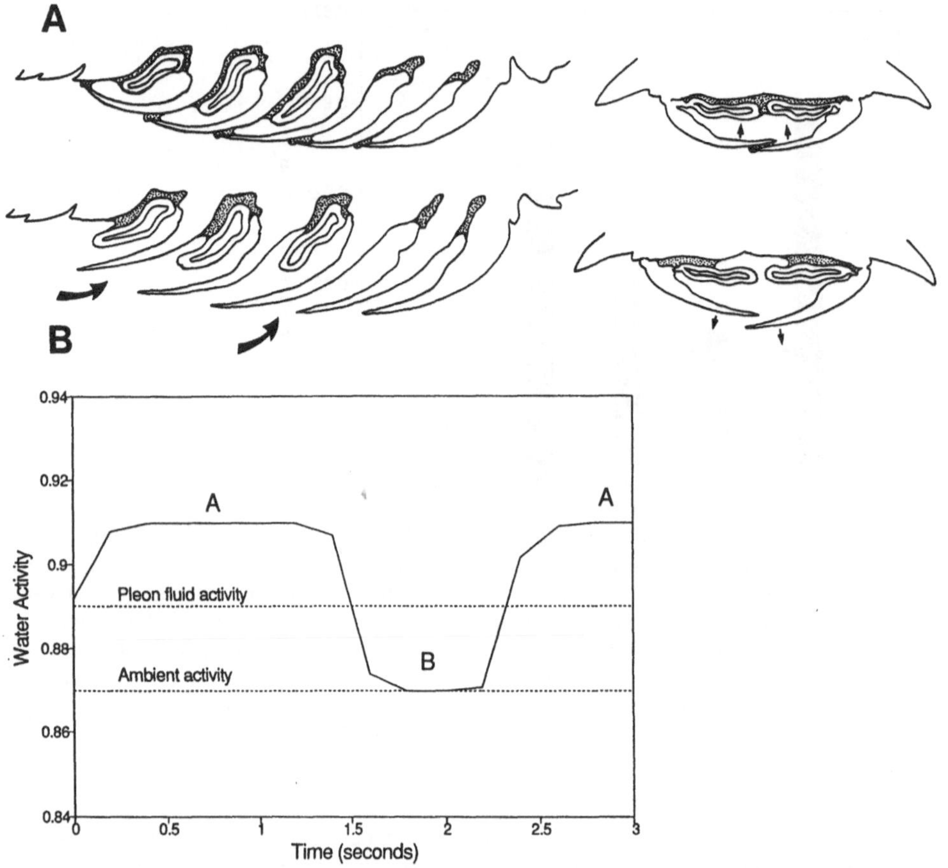

Figure 4. Schematic diagram illustrating the movements of the pleopods during pressure cycling and the accompanying changes in the water activity of air within the PV. Illustrations show off-centre sagittal sections (anterior to right) and right illustrations are of transverse sections, both in the plane of the pleopod-sternite articulations. The sub-pleopodal chambers are contiguous – in a median sagittal section, the PV appears as a single cavity. Pleon fluid activity and ambient activity are assumed to remain constant. In A) Depression of the exopods, and sealing of the PV by marginal pleon fluid (stippled), creates an elevation of pressure and corresponding increase in water activity of the air mass. When this exceeds the pleon fluid activity, water will condense into the pleon fluid. In B) Elevation of the exopods flushes the PV with ambient air ready for another compression cycle. The areas enclosed between the plotted curves for pleon fluid activity and PV air activity during compression (A) and elevation (B) are proportional to the respective gains and losses of water.

the accompanying ventilatory patterns which display increased beat amplitude and peripheral displacement of fluid during pleopodal depression.

The occurrence of pressure cycling in *P. scaber* has been independently confirmed by enclosing the animal within a sealed, Plexiglas chamber and monitoring cyclical pressure reductions in the surrounding air, synchronous with pleopodal depression, by means of a micro-manometer (Wright & Machin 1993b). The decline in ambient pressure during pleopodal depression indicates the sealing and succeeding compression of an enclosed air space, with a corresponding increase in the net vol-

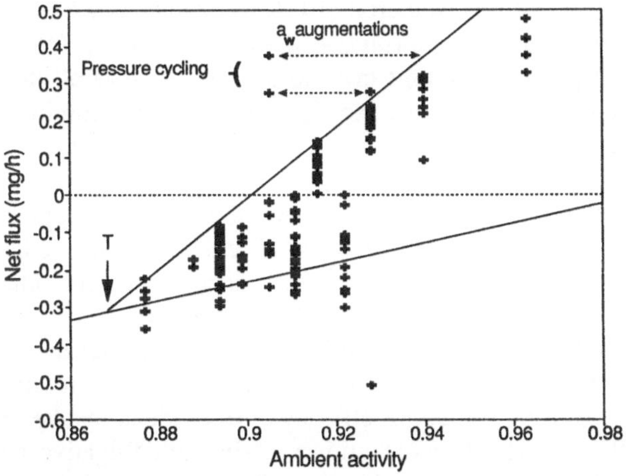

Figure 5. Plot of net flux (mg h^{-1}) against ambient activity, as for Figure 1, but showing 2 examples of proposed pressure cycling (bracketed). Curves were fitted as for Figure 1, including flux data for ambient activities below those shown. During pressure cycling, uptake fluxes are elevated far above the maxima predicted from the linear absorption kinetics. Compression of air within the PV increases its activity above ambient, thus boosting the vapour pressure gradient driving absorption. In the absence of pressure cycling, equal uptake fluxes would only be attainable in much higher ambient activities; the corresponding activity differences represent the 'activity augmentations' generated by compression. (T = uptake threshold). From Wright & Machin (1993b).

ume of the surrounding air space. No net volume changes would occur in the absence of sealing.

Corroborating evidence for occasional interventions of pressure cycling in absorbing animals is provided by gravimetric analyses (Wright & Machin 1993b). Whilst WVA typically conforms to the linear uptake kinetics described above (see Fig. 1), substantially elevated uptake rates are sometimes evident, particularly in ambient activities close to the threshold. We suggest that these anomalously high uptake rates can be explained on the basis of colligative absorption only if the humidity at the uptake site is in fact, higher than that surrounding the whole animal. In other words, the animal appears to have some mechanism of elevating the humdity within the PV. This elevation could be accomplished by a compression of the air by the pleopodal exopods. The increases in ambient humidity generated by compression can be measured directly from plots such as Figure 5, and are relatively consistent across the 7 species of Crinocheta for which the phenomenon has been studied. They range from 0.033 to 0.056, and are equivalent to pressure increases of 3.7 to 6.4 kPa. The resultant increases in the driving force for water uptake are equivalent to osmotic gradients of 1.9 to 3.3 Osm kg^{-1}. *P. scaber*, for example, can generate a fluid osmolality of approximately 7 Osm kg^{-1}, but can achieve uptake rates during pressure cycling that would otherwise require fluid osmolalities of approximately 9.8 Osm kg^{-1}.

1.3 *Comparisons with other vapour-absorbing groups*

The oniscideans constitute the first group of arthropods in which a role of pressure cycling has been confirmed. Compression of an enclosed air space as a means of increasing water activity was first proposed by Maddrell (1971) to explain WVA in the lepismatid *Thermobia domestica* Packard (= *Lepismodes inquilinus*). Corbet (1988) subsequently proposed that such pressure cycling may be utilised by several other groups, as well as constituting an important mechanism for water conservation in insect tracheal systems. It seems probable that many further examples await discovery.

Oniscideans are capable of generating the most rapid vapour-uptake fluxes described to date. When corrected for activity deficits above the threshold, uptake fluxes (μg h^{-1} Pa^{-1}) are 1.3 to 8 times more rapid than those of tenebrionid beetle larvae, and are 1 to 3 orders of magnitude higher than fluxes documented for other groups (Wright & Machin 1993a). Furthermore, if uptake fluxes are corrected for the estimated surface area of the condensing structure (μg h^{-1} cm^{-2} Pa^{-1}), oniscideans absorb between 4 and 100 times more rapidly than other absorbers for which reliable area data are available. The reasons for the superior efficacy of oniscidean uptake are unclear. The rectal complex of tenebrionid larvae utilises a similar colligative absorption mechanism (Machin 1976; O'Donnell & Machin 1992), but ventilation of air within the rectal cavity is limited and purely tidal. In addition to superior ventilation, the Oniscidea may gain a further advantage by their ability to expose hyperosmotic fluid to the humid air directly, avoiding intervening diffusion barriers. Their high uptake fluxes may be seen as an adaptive solution to compensate for modest uptake thresholds and relatively rapid rates of transpiratory water loss. They enable proportional rates of water recovery in saturated air (%fw h^{-1}) comparable to those for *Thermobia* and somewhat higher than for tenebrionids, though as much as an order of magnitude lower than are attainable by the Psocoptera and Mallophaga (see O'Donnell & Machin 1988; Wright & Machin 1993a).

1.4 *Eco-physiological implications of WVA in Oniscidea*

The Oniscideans retain a number of features of their marine ancestors which have been assumed to act as constraints for terrestrial adaptation. These include: retention of the permeable pleopodal endopods, which probably serve as important respiratory epithelia in many species; retention of ammonotely and maxillary urination; and retention of the isopod marsupium and associated viviparity. Consequent physiological limitations for water balance and associated osmotic and ionic homeostasis have been discussed many times (Edney 1954, 1968; Warburg et al. 1984; Wieser 1984) and led to a general acceptance that oniscideans, despite their evident success in diverse ecotypes, possess only moderate physiological adaptations for terrestriality. Such a view has fuelled an interest in the behavioural regulation of water balance by mechanisms which include circadian control of their nocturnal activity patterns, thigmotaxis, and well-described photo-, hygro-, clino- and thermokinetic orienting responses (Cloudsley-Thompson 1952, 1956, 1974, 1977; Sutton 1980; Warburg 1968). Whilst these probably represent the most important components of water-balance regulation in the chiefly endogean Synocheta, the novel finding of WVA in the Crinocheta clearly demands a major rethinking of their water economy.

Oniscideans possess relatively inefficient water barriers by the standards of other terrestrial arthropods (Edney 1977; Hadley & Quinlan 1984). Comparative transpiratory data for 12 common temperate Crinocheta, excluding losses attributable to maxillary urination, reveal standardized fluxes between 0.49 μg h^{-1} cm^{-2} Pa^{-1}

(*Armadillidium vulgare*) and 1.56 μg h^{-1} cm^{-2} Pa^{-1} (*Cylisticus convexus*) (Wright & Machin 1993a). These translate into water-losses of 2.88% fw h^{-1} and 11.11% fw h^{-1}, respectively, for representative-sized animals in dry air at 20°C. Compensatory water sources are obtained from feeding and metabolism. Passive uptake of water vapour across the body surface in saturated air constitutes an additional means of water gain, but is of negligible significance (Wright & Machin 1993a). Data reviewed in Wieser (1984) estimate rates of food absorption in oniscideans to be in the order of 12 mg(dw) $g(fw)^{-1}$ day^{-1} at 20°C. If this is assumed to comprise deciduous litter, with a water content of ca. 60% in an ambient humidity of 80% (Buxton 1924), dietary water intake will be approximately 18 mg $g(fw)^{-1}$ day^{-1} or 0.075% fw h^{-1}. Animals would not be capable of selective augmentation of water absorption since faecal pellets are in passive equilibrium with the haemolymph (Wright & Machin 1993b). Based on a typical resting metabolism for Oniscidea of 200 μl O_2 g^{-1} h^{-1} at 20°C (Wieser 1984), we can calculate that metabolic water would contribute approximately 0.016% fw h^{-1}. Taken together, dietary and metabolic water sources would therefore represent ca. 0.091% fw h^{-1}, and would compensate for only 0.82% to 3.2% of predicted transpiratory losses in dry air. In 50% RH, the transpiratory losses would be approximately halved, in 75% RH they would be reduced 4-fold, and so-on. Thus, in the absence of alternative water-uptake mechanisms, the high permeability of oniscideans would restrict them to humid microhabitats, and would permit only brief foraging in significantly subsaturated conditions. Given that workers normally observe oniscideans during the day, when the animals typically inhabit humid retreats, the hygric implications of both physiological and ecological data have been, for the most part, unchallenged (Cloudsley-Thompson 1977; Edney 1968).

In fact, a considerable body of evidence indicates that Crinocheta regularly display nocturnal activity in appreciably subsaturated humidities (70 to 80% RH) for extended periods (Brereton 1957; Den Boer 1961; Paris 1963). From the data given above, and even given saturated diurnal refuges, this would lead to lethal desiccation of most species over the course of 1 to 3 days. WVA provides a means by which substantial water-debts can be recovered over the course of a few hours. Using comparative data obtained for transpiratory losses and uptake fluxes in different ambient humidities, and assuming a 12h:12h circadian activity pattern, Wright & Machin (1993a) compiled water budget curves for 8 species of Crinocheta. These show that diurnal WVA, in 100% RH, would allow most species to replenish near-lethal levels of water loss (ca. 34% fw) sustained during nocturnal foraging. Correcting these water budget analyses for dietary water intake, we can predict that *A. vulgare* would be capable of routine nocturnal foraging in relative humidities above 29% at 20°C, and even a permeable species such as *C. convexus* could forage in humidities down to ca. 74%. Whether oniscideans regularly follow a pattern of near-maximal levels of water loss and subsequent replenishment remains undetermined. Data from Den Boer (1961) show that a Dutch population of *P. scaber* in aspen woods in was routinely active in humidities down to ca. 70% RH. This is well above the RH which would result in lethal desiccation over a 12-hour foraging period. The water loss sustained in 12 h at 70% RH could be recovered by a subsequent 12h period of WVA in humidities of 96.7% or higher (see Wright & Machin 1993a). Availability of sufficiently humid microhabitats for WVA may, therefore, set a lower limit on the humidities in which animals can sustain nocturnal activity.

These data suggest that adaptations for diurnal and nocturnal water balance merit independent consideration. *P. spinicornis* has a moderate integumental permeability (standardized flux = 0.69 μg h^{-1} cm^{-2} Pa^{-1}), and would sustain lethal desiccation in humidities below 46% RH for a 12h nocturnal foraging period. However, it is a rapid vapour absorber, theoretically capable of replenishing maximum tolerable water losses in a mean diurnal humidity of 97.5% given 12h inactivity. By comparison, *Oniscus asellus* would be limited to mean foraging humidities above 69% for a 12h night, and would require more than 12h to replenish maximum tolerable losses by WVA, even in saturated air. *P. spinicornis* could thus be classed as a mesic nocturnal forager, but a xeric diurnal 'rester', whilst *O. asellus* is clearly a hygric species by both nocturnal and diurnal classifications. These theoretical predictions from physiological data are well supported by results of ecological surveys (Hatchett 1947; Vandel 1962; Harding & Sutton 1985).

1.5 *WVA in a diplochete, Ligia oceanica*

The somewhat surprising discovery of WVA in the supra-littoral isopod *Ligia oceanica* reveals a physiological mechanism by which this species can attain considerable independence from intertidal, liquid water sources. Although capable of prolonged hyper-regulation in aerated 25% seawater, and tolerating indefinite submergence in 100% seawater (Parry 1953; Todd 1963), *L. oceanica* may also occur on cliffs and in coastal grasslands up to 150 m above mean sea level (Harding & Sutton 1985). The 'open' marsupium, requiring fluid renewal from an external water source (Hoese 1984), may constitute the chief factor limiting terrestrial diversification in this species. The relatively high transpiratory water losses (8.51% fw h^{-1} in dry air), and a relatively high critical equilibrium activity for WVA (0.937) would restrict *L. oceanica* to high humidities (Wright & Machin 1993a). However, water losses up to ca. 20% could be replenished over a 12-hour period in saturated air.

1.6 *Water balance in Synocheta.*

The apparent absence of WVA in the Synocheta, including the familiar Trichoniscidae, may explain their generally cryptozoic habits and inability to invade xeric niches. The small size of most species (<5 mm), combined with high integumental permeabilities, results in water-loss rates of ca. 40% to 60% fw h^{-1} in dry air at 20°C (Wright & Machin 1993a). Over a 12-hour foraging period, this would result in lethal desiccation in humidities below 92 to 94%. Water losses can only be renewed from dietary and metabolic sources. Assuming the values quoted above, these sources would provide a mean water gain of only 0.091% fw h^{-1}. Transpiratory water losses will exceed this value in any humidity below 98.8% RH, (a_w = 0.988) assuming a haemolymph a_w of 0.99. It is thus clear, on the basis of available data, that Synocheta have only very limited capacity to exploit sub-saturated environments.

2 AMMONOTELY IN TERRESTRIAL ISOPODS

Oniscideans constitute an exception to 'Needham's Rule' (Needham 1938) that terrestriality is accompanied by a shift from ammonotely to water-conserving ureotelic and purinotelic forms of nitrogenous excretion. Their form of ammonotely is unusual, however, in that ammonia is liberated primarily as a gas. Evidence for elimination of gaseous ammonia in terrestrial isopods was first provided by Dresel & Moyle (1950). Although their measurements indicated that only 10% to 30% of total ammonia was volatilized, W.C. Sloan suggested subsequently (unpublished observations cited by Hartenstein 1968) that the major fraction of ammonia was eliminated in gaseous form. This view has been vindicated by quantitative analysis of nitrogenous end products (Hartenstein 1968; Wieser & Schweizer 1970). Ammonia appears to comprise about 95% of total nitrogenous waste, and most is volatilized; less than 10% is voided in soluble form with the faeces.

Liberation of ammonia shows a pronounced diel rhythm in fasting animals. By collecting gaseous ammonia in acid traps, Wieser et al. (1969) and Kirby & Harbaugh (1974) showed that most ammonia is eliminated diurnally when animals are inactive. The latter authors also identified a similar diel rhythm in ammonia excretion by three aquatic marine isopods from the family Valvifera; two species of Flabellifera showed an approximate reversal of this pattern, with an excretory maximum between 0400 and 0800 h. Wieser et al. (1969) and Wieser & Schweizer (1970) showed that ammonia volatilization in feeding *P. scaber* is reduced 3 to 4-fold, presumably due to reduced catabolism of tissue proteins; the pronounced diel rhythm evident in fasting animals was almost eliminated in feeding animals. Furthermore, these workers demonstrated a pronounced seasonal variation in total N output of *Oniscus asellus* and *P. scaber*, with spring/summer production approximately 5-fold higher than that in autumn/winter. This variation is consistent with the greater availability of deciduous litter in the latter part of the year.

In the short term, ammonia volatilization is discontinuous (Wieser et al. 1969; Wieser & Schweizer 1970; Wieser 1972a). Some fasting animals released only one burst of ammonia in 24 days, whereas others produced one or more bursts per day (Wieser et al. 1969). These studies are based on ammonia collected in acid traps over the course of several hours. Incidences of ammonia release in 30-minute intervals were investigated by Kirby & Harbaugh (1974) by wrapping individual *P. scaber* in phenol red indicator paper and revealed bouts of liberation restricted to within the 30-minute sampling period, as well as bouts extending over 3 successive sampling periods. Similar discontinuities within the diurnal excretory period were established for the valviferan *Idotea resecata*. This contradicts the generally held belief that ammonia excretion in aquatic Crustacea is a continuous process (Regnault 1987). However, more accurate time courses of ammonia release over the short-term await further study.

Two mechanisms have been suggested to explain the volatilization of ammonia. Wieser (1972a) proposed that ammonia is exchanged between the haemolymph and pleon fluid, from which it could be volatilized by alkalinization. Increased pH would promote dissociation of the weakly acidic NH_4^+, yielding the volatile base NH_3; since ammonia has a pK of ca. 9.25, it follows from the Henderson-Hasselbalch equation that the NH_3/NH_4^+ ratio will equal unity at this pH. At lower pH's, closer to the physiological range, the concentration of NH_4^+ will exceed that of NH_3 by 50 to 100-fold. Hoese (1981) suggested an alternative excretory pathway, with ammonia being released across the epithelia of the paired maxillary glands, although he also proposed alkalinization in the PV as a mechanism for volatilizing NH_3. On the basis of SEM studies and dye injections he demonstrated that maxillary urine is conveyed posteriorly via the water-transport system or 'Wasserleitungssystem' (WS) which forms a capillary conducting channel along the lateral pleurae. In addition to alkalinization, Hoese proposed that the PV functions in solute resorption, with remaining fluid being resorbed in the rectum. It is worth pointing out that the two proposed mechanisms need not be mutually exclusive: both oral and pleoventral sites of ammonia release were identified by Kirby & Harbaugh (*ibid.*) using phenol red indicator paper. Although a higher incidence of pleoventral release was observed, these studies do not permit a quantitative evaluation of the proposed excretory pathways.

Measurements of total ammonia concentrations (NH_3 + NH_4^+) in body fluids of *P. scaber* indicate temporal variability comparable to that observed in excretory rates (Wright & O'Donnell 1993). Whilst concentrations are usually below 1.5 mmol l^{-1}, animals periodically display dramatically elevated levels of ammonia in the haemolymph, with concentrations frequently exceeding 20 mmol l^{-1} and exceptionally exceeding 50 mmol l^{-1} (Fig. 6). Measured durations of such increases vary between approximately 20 min and 2h. There are synchronous increases in pleon fluid ammonia levels. Ammonia increases are significantly correlated with ventilatory movements of the pleopods, such as are observed during WVA. These findings are compatible with the periodic release of ammonia into the haemolymph, and its subsequent movement into the pleon fluid from which NH_3 may be volatilized. Whether ammonia diffuses from haemolymph to pleon fluid as NH_3, or whether NH_4^+ diffuses outwards, perhaps in exchange for Na^+, remains unclear. Some bouts of maxillary urination coincide with these increases in haemolymph ammonia, and urine ammonia

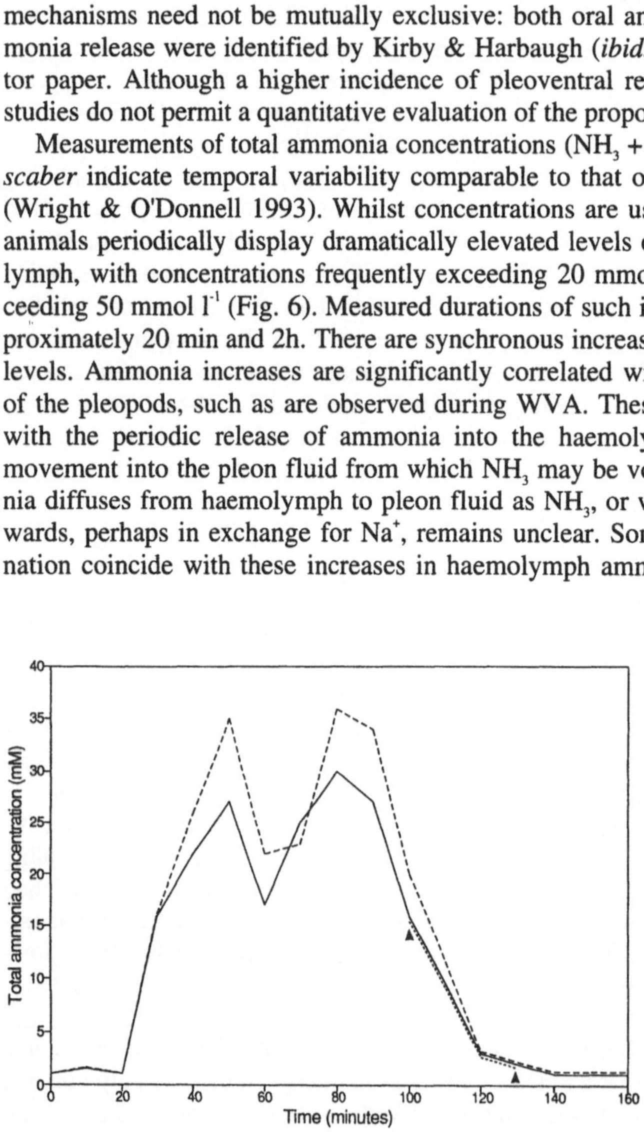

Figure 6. Semi-schematic diagram, based on Wright & O'Donnell (1993), illustrating the relationship between total ammonia levels in different body fluids of an individual *P. scaber*. The plot shows a bout of assumed ammonia volatilization, characterized by a sharp increase in haemolymph total ammonia concentrations (solid line) and concomitant increase in pleon fluid concentrations (dashed line). Pleon fluid concentrations are typically somewhat higher than those in haemolymph. Maxillary urine (dotted line) may be excreted any time – the duration of this excretory bout is marked by arrows – and reveals ammonia concentrations isosmotic with those in haemolymph.

levels remain equimolar with haemolymph levels. These urinations will inevitably contribute to the liberation of volatile NH_3. However, the absence of any correlation between the timing of periodic haemolymph ammonia elevations and bouts of maxillary urination indicate that urine is of only incidental importance in ammonotely.

Recent data do not support any role of alkalinization in increasing the NH_3/NH_4^+ ratio and thereby accelerating volatilization. Both haemolymph and pleon fluid pH values vary between approximately 7.5 and 7.6 whether in ventilating, pre-ventilating (following initial secretion of uptake fluid prior to WVA) or non-ventilating animals (Wright & O'Donnell 1993). Urine pH is in the range 7.1-7.8. These values were derived from *in situ* measurements using H^+-selective microelectrodes.

In addition to the proposed volatilization of NH_3 from the pleon fluid, a certain amount of NH_3 will inevitably be lost by transpiration across the general cuticle. Rates of diffusion of NH_3 across the body wall can be predicted from estimates of mean activity gradients for NH_3 diffusion, and assuming that the diffusion coefficient of NH_3 across isopod cuticle will be similar to that for water (see Wright & O'Donnell 1993). Such estimates suggest that passive cutaneous losses could account for only 10-20% of NH_3 volatilization in fasting *P. scaber*, based on the long-term excretion rates documented by Wieser (1984). However, the 3 to 4-fold reduction in ammonia release in feeding relative to starved animals may leave cutaneous diffusion as a more significant component of volatilization, and may explain the virtual disappearance of the diel rhythm in feeding animals. This point is discussed in more detail below. Passive cutaneous loss of NH_3 in less permeable species, such as *Venezillo arizonicus* and *Hemilepistus afghanicus* (see Coenen-Stass 1981), would be negligible, if one assumes similar mean haemolymph ammonia levels (and thus similar partial pressure gradients for diffusion) to those measured for *P. scaber*.

2.1 *Ammonia toxicity and sequestration*

Maximum concentrations of haemolymph ammonia in *P. scaber* exceed the intracellular levels tolerated in vertebrates more than 100-fold (Lehninger 1982, Meijer et al. 1990). High extracellular ammonia levels have been reported in several other Crustacea (Greenaway 1991). A mean haemolymph concentration of 20.0 mmol l^{-1} has been measured for *Uca pugilator* (Green et al. 1959) and concentrations as high as 271 mmol l^{-1} and 131 mmol l^{-1}, respectively, have been determined for the oplophorid shrimps *Notostomus gibbosus* and *N. elegans* (Sanders & Childress 1988). During dehydration, the gecarcinid *Cardiosoma carnifex* tolerates sustained haemolymph ammonia concentrations of 7 mmol l^{-1} (Wood et al. 1986). Taken together, these studies raise the possibility that tolerance of elevated ammonia levels is a common feature of Crustaceans. Wieser & Schweizer (1972) have pointed out that high ammonia levels in the integument of isopods (and other crustaceans) will favour dissociation of bicarbonate, thus promoting the deposition of $CaCO_3$. Volatilization of NH_3 from the mantle wall of pulmonates (Speeg & Campbell 1968) is thought to serve in a similar role during shell growth (Edney 1977).

Much less is known about intracellullar ammonia levels in isopods and other crustaceans. Hartenstein (1968) and Wieser & Schweizer (1972) reported ammonia concentrations up to 17 mmol l^{-1} in the somatic tissues ('body wall') of *P. scaber* and

O. asellus, although their samples are likely to have included some haemolymph. Nevertheless, such data suggest likely tolerance of high intracellular ammonia levels, at least in the short term. The toxic effects of ammonia include inhibition of the Krebs cycle through competitive depletion of ketoglutarate, disruptions of chloride channels and hence neuronal hyperpolarization, as well as perturbations of the metabolism of neurotransmitters, notably glutamate and aspartate. High levels of intracellular ammonia may be tolerable if the molecule is sequestered in specific membrane-delimited vesicles, as is the case in chaetognaths, for example (Bone et al. 1987). However, no comparable vesicles which could fulfil such a role have been observed in isopod tissues. An alternative possibility is that a carrier molecule transports ammonia in the haemolymph, so that freely diffusible levels of the molecule are less than the total concentration.

The periodic bouts of elevated haemolymph ammonia concentrations observed in *P. scaber* indicate the presence of a non-toxic precursor from which ammonia can be released by deamination. In Crustacea, deamination by a dehydratase has only been demonstrated for L-glutamate, serine, proline and perhaps threonine (Fellows & Hird 1979; Schoffeniels 1976, 1984). Amino groups may also be sequestered as amines, such as glutamine and asparigine (Regnault 1987). Glutamine and glutamate are both present in high concentrations (up to 11.5 and 15.7 mM l^{-1} respectively) in the somatic tissues of *P. scaber* (Wieser & Schweizer 1972). Glutamine is also present in high concentrations (>500 μmol l^{-1}) in the haemolymph of terrestrial oniscideans (Sevilla & Lagarrigue 1974), although the haemocyte/plasma ratios are unknown. We have recently shown that when *P. scaber* is exposed to NH_3 in the surrounding atmosphere, whole body glutamine and glutamate levels increase in proportion to the duration of exposure and the pNH_3. Ammonia sequestered as Glu and Gln is subsequently volatilized when animals are transferred to a humid, ammonia-free atmosphere (Wright et al. 1994). Approximately 36% of the nitrogen excreted as NH_3 can be accounted for by deamination of Glu and Gln. Subsequent analyses of amino acid levels following ammonia loading of animals in high pNH_3 have confirmed glutamine, arginine and glycine as the major forms of sequestered ammonia (Wright et al. 1994).

Further support for a model incorporating somatic synthesis of glutamine and periodic release of ammonia into the haemolymph is provided by determinations of elevated glutaminase activity in the somatic tissues (Wieser & Schweizer 1972; Wieser 1972b). Furthermore, glutaminase activity displays seasonal variation (Wieser & Schweizer 1972) which concurs with the seasonal pattern of ammonia production. Maximal glutaminase activity occurs in the pH range 7.7-8.0; the enzyme is not present in soluble form, and may occur in the mitochondria.

Evidence for a primary role of glutamine in ammonia sequestration is also available for other Crustacea. The semi-terrestrial crab *Paratelphusa hydrodromus* shows a decline in glutaminase activity as ammonotely is substituted by ureotely in increasing salinities (Krishnamoorthy & Srihari 1973). Elucidation of ammoniogenesis pathways in the crabs *Carcinus* and *Cancer* also indicates glutamine as the major storage product in these species (King et al. 1985). Elevated glutaminase activity in the gills suggests that branchial excretion predominates.

2.2 *Coupling of ammonia volatilization and WVA.*

Several lines of evidence suggest that ammonia excretion and WVA are physiologically coupled. Both processes are associated with pleon fluid secretion and pleopodal ventilation. As well as providing circulation of humid air for WVA, ventilation would dispel ammonia, thereby serving adaptive functions in both processes. Furthermore, we have not observed periodic elevations of haemolymph ammonia when WVA is precluded by exposing isopods to humidities below the uptake threshold (Wright & O'Donnell 1993). It seems probable that the clear diurnal maxima for ammonia release in fasting animals coincide with periods of vapour uptake, during which animals may replenish water losses sustained during nocturnal foraging (Wright & Machin 1993a). Similarly, the disappearance of a clear diel rhythm in the reduced NH_3 volatilization of feeding animals may be due to a predominant fraction diffusing across the cuticle, thereby uncoupling ammonia release from diurnal WVA. Studies comparing haemolyph ammonia levels in fasting and feeding isopod populations would be interesting in this regard. If feeding populations excrete a large fraction of their ammonia by passive cutaneous diffusion, this would obviate the need for sequestration and much lower glutamine and glutamate levels might therefore be predicted.

Preliminary studies with *P. scaber* have demonstrated substantial apical effluxes of ammonium from excised endopods exposed to high ambient ammonia levels. These studies involved bathing both basal and apical surfaces with saline containing elevated NH_4Cl concentrations (Wright et al., unpublished data). Effluxes were calculated from ammonium concentration gradients measured by an oscillating ammonium-selective microelectrode. Automated movements of the microelectrode at ca. 0.8 Hz in an axis normal to the apical membrane allowed concentrations to be measured at the two extremes of the excursion (Kuhtreiber & Jaffe 1990). Although the transport mechanism is metabolically dependent, the form and pathway of transport (transcellular or paracellular) await determination. NH_4^+ secretion may involve apical Na^+/NH_4^+ exchange, exploiting the large inward diffusion gradient for Na^+ present during WVA. However, if Na^+/NH_4^+ exchange does occur, it is apparently not amiloride-sensitive.

Whether WVA and NH_3 volatilization are physiologically inter-dependent processes, or merely facultatively associated, their apparent temporal association would allow N-excretion to be coupled to a net water gain. This would apply in any ambient activity exceeding the critical equilibrium activity (0.913 for *P. scaber*); in contrast, ammonia volatilization in the absence of WVA would lead to simultaneous water losses in any ambient activity below that of the haemolymph (ca. 0.988). So far as we are aware, such a coupling of N-excretion to a significant water gain would be unique in terrestrial animals. Furthermore, NH_3 volatilization during WVA would not compromise water uptake: if NH_3 is transferred to the pleon fluid via an antiport mechanism (Na^+/NH_4^+ exchange), the osmolality of the uptake fluid will be unaltered; if excretion involves passive diffusion of NH_3, the osmolality of the pleon fluid may actually increase by a small amount.

It seems likely that colligative water vapour absorption would be preadaptive for facultative ammonia volatilization in other terrestrial arthropods. *Tenebrio,* for example, could volatilize NH_3 from the rectal wall during colligative vapour uptake (see Machin 1980), as may lepismatids and flea larvae (see O'Donnell & Machin

1988), although colligative mechanisms in these latter groups are undetermined. Excretion of ammonia (as NH_3 or NH_4^+) may also be coupled to water resorption in the recta of other, non-vapour-absorbing insects. Both *Schistocerca* and *Periplaneta* are capable of dehydrating the faeces (Loveridge 1974). *Schistocerca* oxidises amino acids in the hindgut and secretes NH_4^+ into the lumen (Thomson et al. 1988). In unfed locusts, NH_4^+ constitutes ca. 40% of excreted nitrogen, and is voided in the acidic urinary pellets as soluble NH_4Cl, and as insoluble ammonium phosphate and ammonium urate (Harrison & Phillips 1992). Similar excretory products may be utilized by *Periplaneta,* which excretes over 50% of nitrogenous waste as faecal/urinary ammonia (Mullins 1974; Mullins & Cochran 1973, 1974). In other species, low rectal pH could favour diffusion trapping of NH_3 as NH_4^+, reducing back-diffusion into the haemolymph (Towle 1990).

3 ORIGINS OF TERRESTRIALITY IN ONISCIDEA

There is general agreement that the terrestrial oniscidean isopods evolved from marine ancestors (Little 1983). The Diplocheta are predominantly intertidal and show only minor modifications of the marsupium compared to marine isopods. By comparison, the Crinocheta have evolved a 'closed' marsupium and closed water transport system (WS) (Hoese 1981, 1984). A marine origin for oniscideans is further suggested by their relatively high haemolymph osmolalities, and haemolymph electrolyte compositions proportionally similar to that of sea water (Little 1983).

Information is scant concerning ammonotely or possible WVA in the Tylidae. Vandel (1965) considered them to have colonised land independently of other oniscideans. Recent study of *Helleria brevicornis* reveal a lack of WVA and fecal elimination of ammonia as salts (Wright unpubl.), which tends to support Vandel. So far as the remaining groups are concerned, the most parsimonious phylogenies would derive the Crinocheta from vapour-absorbing Diplocheta. Depending on the distribution of WVA within the latter section, the Synocheta may be derived from a non-absorbing lineage of the Diplocheta, or may constitute a sister-group. The ecologic diversity of extant Synocheta strongly suggests that they have never undergone a significant adaptive radiation from endogean (interstitial) habitats. This inability to colonise mesic/xeric habitats may relate to the strong selection pressure for small size in interstitial organisms. As a result, unfavourable surface area: volume ratios compromise water conservation in desiccating conditions.

A detailed discussion of the origins of WVA is premature, but the identification of vapour uptake in *Ligia oceanica* warrants a brief mention in this context. Prior to the discovery of WVA in isopods, vapour absorption was believed to be associated primarily with xeric groups. Oniscideans constitute an exception to this generalization, being predominantly mesic-hygric, even within the Crinocheta. We suggest that the evolution of WVA in terrestrial isopods has resulted from two primary factors: 1) Strong, directional selection pressures to counter substantial evaporative water losses, the latter attributable primarily to relatively permeable integuments; and 2) Strong preadaptations for the evolution of a colligative vapour absorption mechanism.

Water-stress is generally associated with a water-poor diet, as for mallophagans or grain-feeding anobiid and tenebrionid beetle larvae, or with xeric habitats, as for desert insects including *Arenivaga* and many lepismatids (O'Donnell & Machin 1988). However, animals living in mesic/hygric habitats may also experience appreciable water stress, since their integuments may be relatively permeable. There is thus no *a priori* reason to assume that selection pressures for WVA in terrestrial isopods have been less pronounced than in vapour-absorbing insects and arachnids. The scarcity of WVA among flying insects – the only documented examples belonging to the Psocoptera (Rudolph 1982) – illustrates the overwhelming advantage of flight for behavioural evasion of water-stress.

Like the water-resorbing hindguts of many insects, the salt-secreting pleon of hyporegulating isopods is elegantly preadapted for colligative WVA. The predominantly intertidal habitats of *Ligia* spp. support a littoral origin for vapour absorption; there is no indication that the Ligiidae are more advanced terrestrial forms which have secondarily reinvaded littoral niches (Edney 1968; Hoese 1981, 1984; Little 1983). Their intertidal habits do not, however, suggest desiccation as the primary selection pressure underlying the evolution of WVA. The majority of Ligiidae are large (>50 mg) and could withstand 12-hourly intertidal emersion in humidities above ca. 80%, as well as being able to tolerate prolonged submergence within wide osmotic fluctuations (Parry 1953; Todd 1963; Wilson 1970). *L. pallasii and L. occidentalis* are efficient hyporegulators (Wilson 1970), and would be capable of exploiting hypersaline pools for water replenishment when desiccated. Littoral Ligiidae are also subject to dietary salt-loading, their diet comprising shore debris, including algae and halophytes. Hyporegulation requires the excretion of Na^+ and Cl^-, presumably from the endopods. Emersed animals could secrete salts into the PV and periodically release them in concentrated solution via the apposed uropods. Oniscideans are well known to release excess fluid from the PV and WS via uropodal and pereopodal capillary channels (Spencer & Edney 1954; Hoese 1981; Wright & O'Donnell, personal observations).

Accumulation of concentrated NaCl in the PV would generate colligative vapour condensation in ambient activities above that of the fluid. Combined with the evolution of the WS and posterior recycling of maxillary urine (Hoese 1981), colligative vapour condensation could thus be exploited for water uptake, with salt only intermittently excreted. In this way, *Ligia* spp. could vary rates of salt excretion in relation to water uptake to offset hyperosmosis induced by salt-loading and/or desiccation. This would present an obvious preadaptation for terrestrial colonization where salts in general would be conserved, but their secretion and resorption in the PV would be retained for colligative WVA to counter desiccation.

Rapid pleopodal ventilation (>2 Hz) is a continuous process in aquatic isopods, circulating a respiratory current over the endopods, and is equally apparent in terrestrial species that become inadvertently submerged. In oniscideans, the inherited neuro-muscular control for ventilation permits the slower ventilatory movements which circulate uptake fluid and maxillary urine during posterior recycling. Oniscideans also display increasingly effective sealing of the PV by the pleopodal exopods, isolating the permeable endopods from desiccating conditions (Unwin 1932; Edney & Spencer 1955). Together, ventilation and close apposition of the exopods and pleon sternites provide a preadaptive basis for pressure cycling. Whether pressure cycling is utilised by *Ligia oceanica* is currently unknown. Further studies on the

Diplocheta in relation to haemolymph osmoregulation on land, WVA, and incidences of pressure cycling, may throw further light on the evolution of WVA in isopods.

The evolution of gaseous ammonia liberation in terrestial isopods, as with other terrestrial crustaceans, gastropods and annelids, probably involved very little adaptive modification of existing excretory organs. The permeable endopods probably serve as the major sites for ammonia exchange in both aquatic and terrestrial families. The demonstration of diurnal maxima of ammonia release in marine as well as terrestrial species (Kirby & Harbaugh 1974) may reflect, in part, similarities in neuroendocrine control but these, in turn, could be regulated by a variety of environmental and physiological variables (see Regnault 1987). The intermittent release of strongly elevated concentrations of ammonia into the haemolymph of *P. scaber* requires confirmation in other oniscideans. However, it clearly suggests an adaptation to reduce desiccation, restricting bouts of ammonia release to short periods in which elevated concentrations will accelerate volatilization.

REFERENCES

Bone, Q., C. Brownlee, G.W. Bryan, P.R. Dando, M.I. Liddicoat, A.L. Pulsford & K.P. Ryan. 1987. On the differences between the two 'indicator' species of chaetognath *Sagitta setosa* and *S. elegans*. *J. Mar. Biol. Ass. U.K.* 67: 545-560.

Brereton, J. Le G. 1957. The distribution of woodland isopods. *Oikos* 8: 85-106.

Buxton, P.A. 1924. Heat, moisture and animal life in deserts. *Proc. R. Soc. Lond.* (B)96: 123-131.

Cloudsley-Thompson, J.L. 1952. Studies in diurnal rhythms II. Changes in the physiological responses of the woodlouse *Oniscus asellus* to environmental stimuli. *J. Exp. Biol.* 29: 295-303.

Cloudsley-Thompson, J.L. 1956. Studies in diurnal rhythms VII. Humidity responses and nocturnal activity in woodlice (Isopoda). *J. Exp. Biol.* 33: 576-582.

Cloudsley-Thompson, J.L. 1974. Climatic effect affecting the nocturnal emergence of woodlice and other arthropods. *Ent. Mon. Mag.* 109: 123-124.

Cloudsley-Thompson, J.L. 1977. *The water and temperature relations of woodlice*. Shildon: Meadowfield Press Ltd.

Coenen-Stass, D. 1981. Some aspects of the water balance of two desert woodlice *Hemilepistus afghanicus* and *Hemilepistus reaumuri* (Oniscoidea, Isopoda, Crustacea). *Comp. Biochem. Physiol.* 71: 405-419.

Coenen-Stass, D. 1989. Transpiration, vapour absorption and cuticular permeability in woodlice (Oniscidea, Isopoda, Crustacea). Pages 253-267 in *Water transport in Biological Membranes, vol. II, From Cells to Multicellular Barrier Systems* (G. Benga, ed.). Boca Raton, Florida: CRC Press, Inc.

Corbet, S.A. 1988. Pressure cycling and the water economy of insects. *Review. Phil. Trans. R. Soc. Lond.* (B)318: 377-407.

Dresel, I.B. & V. Moyle. 1950. Nitrogenous excretion of amphipods and isopods. *J. Exp. Biol.* 27: 210-225.

Edney, E.B. 1954. Woodlice and the land habitat. *Biol. Rev.* 29: 185-219.

Edney, E.B. 1960. Terrestrial adaptations. pp 367-393 in *The Physiology of Crustacea*, T.H. Waterman, ed. Academic Press, New York.

Edney, E.B. 1966. Absorption of water vapor from unsaturated air by *Arenivaga* sp. (Polyphagidae, Dictyoptera). *Comp. Biochem. Physiol.* 19: 387-408.

Edney, E.B. 1968. The transition from water to land in isopod Crustacea. *Amer. Zool.* 8: 309-326.

Edney, E.B. 1977. *Water balance in land arthropods*. Springer-Verlag, Berlin.

Edney, E.B. & J. Spencer. 1955. Cutaneous respiration in woodlice. *J. Exp. Biol.* 32: 256-269.

Fellows, F.C.I. & F.J.R. Hird. 1979. Nitrogen metabolism and excretion in the freshwater crayfish *Cherax destructor*. *Comp. Biochem. Physiol.* 64: 235-238.

Green, J.W., M. Harsch, L. Barr & C.L. Prosser 1959. The regulation of water and salt by the fiddler crabs *Uca pugnax* and *Uca pugilator*. *Biol. Bull.* 116: 76-87.

Greenaway, P. 1991. Nitrogenous excretion in aquatic and terrestrial crustaceans. *Mem. Queensland Mus.* 31: 215-227.

Hadley, N.F. & M.C. Quinlan. 1984. Cuticular transpiration in the isopod *Porcellio laevis*: chemical and morphological factors involved in its control. *Symp. Zool. Soc. Lond.* 53: 97-107.

Harding, P.T. & S.L. Sutton. 1985. *Woodlice in Britain and Ireland: Distribution and Habitat.* Natural Environment Research Council, Institute of Terrestrial Ecology, Huntingdon.

Harrison, J.F. & J.E. Phillips. 1992. Recovery from acute haemolymph acidosis in unfed locusts: II. The role of renal acid and nitrogen excretion. *J. Exp. Biol.* 165: 97-108.

Hartenstein, R. 1968. Nitrogen metabolism in the terrestrial isopod *Oniscus asellus*. *Am. Zool.* 8: 507-519.

Hatchett, S.P. 1947. Biology of the Isopoda of Michigan. *Ecol. Monogr.* 17: 47-79.

Hoese, B. 1981. Morphologie und Funktion des Wasserleitungssystems der terrestrichen Isopoden. *Zoomorphol.* 98: 135-167.

Hoese, B. 1984. The marsupium in terrestrial isopods. *Symp. Zool. Soc. London* 53: 65-75.

King, F.D., T.L. Cucci & R.R. Bidigare. 1985. A pathway of nitrogen metabolism in marine decapod crabs. *Comp. Biochem. Physiol.* 80B: 401-403.

Kirby, P. & R.D. Harbaugh. 1974. Diurnal patterns of ammonia release in marine and terrestrial isopods. *Comp. Biochem. Physiol.* 47A: 1313-1321.

Krishnamoorthy, R.V. & K. Srihari. 1973. Changes in excretory patterns of the freshwater field crab *Paratelphusa hydrodromus* on adaptation to higher salinities. *Marine Biology* 21: 341-348.

Kuhtreiber, W.M. & Jaffe, L.F. 1990. Detection of extracellular calcium gradients with a calcium-specific vibrating electrode. *J. Cell Biol.* 110: 1565-1573.

Kümmel, G. 1981. Fine-structural indications of an osmoregulatory function of the 'gills' in terrestrial isopods (Crustacea, Oniscoidea). *Cell Tissue Res.* 214: 663-666.

Kümmel, G. 1984. Fine-structural investigations of the pleopodal endopods of terrestrial isopods with some remarks on their function. *Symp. Zool. Soc. London* 53: 77-93.

Lehninger, A.L. 1982. *Principles of Biochemistry*. New York: Worth.

Little, C. 1983. *The Colonisation of Land.* Cambridge University Press.

Loveridge, J.P. 1974. Studies on the water relations of adult locusts. II. Water gain in the food and loss in the faeces. *Trans. Rhodesia Sci. Assoc.* 56: 1-30.

Machin, J. 1976. Passive exchange during water vapour absorption in mealworms (*Tenebrio molitor*): a new approach to studying the phenomenon. *J. Exp. Biol.* 65: 603-615.

Maddrell, S.H.P. 1971. The mechanisms of insect excretiory systems. *Advances in Insect Physiology* 8: 19-331.

Meijer, A.J., W.H. Lamers & R.A.F.M. Chamuleau. 1990. Nitrogen metabolism and ornithine cycle function. *Physiol. Rev* 70(3): 701-748.

Mullins, D.E. 1974. Nitrogen metabolism in the American cckroach: an examination of whole-body ammonium and other cations excreted in relation to water requirements. *J. Exp. Biol.* 61: 541-556.

Mullins, D.E. & D.G. Cochran. 1973. Nitrogenous excretory material from the American cockroach. *J. Insect Physiol.* 19: 1007-1019.

Mullins, D.E. & D.G. Cochran. 1974. Nitrogen metabolism in the American cockroach: an examination of whole-body and fat-body regulation of cations in response to nitrogen balance. *J. Exp. Biol.* 61: 557-570.

Needham, J. 1938. Contributions of chemical physiology to the problems of reversibility in evolution. *Biol. Rev.* 13: 225-251.

O'Donnell, M.J. & J. Machin. 1988. Water vapor absorption by terrestrial organisms. PP 47-107 in *Advances in Comparative and Environmental Physiology* Vol. 2. Springer-Verlag, Berlin.

Paris, O.H. 1963. The ecology of *Armadillidium vulgare* in California grassland: food, enemies and weather. *Ecol. Monog.* 33: 1-22.

Parry, G. 1953. Osmotic and ionic regulation in the the isopod crustacean *Ligia oceanica*. *J. Exp. Biol.* 30: 567-574.

Regnault, M. 1987. Nitrogen excretion in marine and freshwater Crustacea. *Biol. Rev.* 62: 1-24.
Rudolph, D. 1982. Occurrence, properties and biological implications of the active uptake of water vapor from the atmosphere in Psocoptera. *J. Insect Physiol.* 28: 111-121.
Sanders, N.K. & J.J. Childress. 1988. Ion replacement as a buoyancy mechanism in a pelagic deep-sea crustacean. *J. Exp. Biol.* 138: 337-343.
Schoffeniels, E. 1976. Adaptations with respect to salinity. *Biochemical Society Symposia* 41: 179-204.
Schoffeniels, E. 1984. Le metabolisme energetique des crustaces euryhalins. In *Biochimie Compare. Collection de Biologie Evolutive*, vol. 9 (ed. E. Schoffeniels), pp 123-131. Masson, Paris.
Sevilla, C. & J.-G. Lagarrigue. 1974. Acides amines libres de l'haemolymphe de *Ligia italica, Porcellio laevis, Armadillidium vulgare et Armadillo officinalis* (Crustaces, Isopodes). C.R. *Acad. Sc. Paris* (D) 278: 1079-1082.
Speeg, K.V. & J.W. Campbell. 1968. Purine biosynthesis and excretion in *Otala (= Helix) lactea*: an evaluation of the nitrogen excretory potential. *Comp. Biochem. Physiol.* 26: 579-595.
Spencer, J.O. & E.B. Edney. 1954. The absorption of water by woodlice. *J. Exp. Biol.* 31: 491-496.
Sutton, S.L. 1980. The population dynamics of *Trichoniscus pusillus* and *Philoscia muscorum* (Crustacea, Oniscoidea) in limestone grassland. *J. Anim. Ecol.* 37 425-444.
Sutton, S.L., M. Hassall, R. Willows, R.C. Davis, A. Grundy & K.D. Sunderland. 1984. Life histories of terrestrial isopods: a study of intra and inter-specific variation. *Symp. Zool. Soc. London* 53: 269-294.
Thomson, R.B., J.M. Thomson & J.E. Phillips. 1988. NH_4^+ transport in acid-secreting insect epithelium. *Am. J. Physiol.* 254: R348-R356.
Todd, M.E. 1963. Osmoregulation in *Ligia oceanica* and *Idotea granulosa. J. Exp. Biol.* 40: 381-392.
Towle, D.W. 1990. Sodium transport systems in gills. *Comp. Physiol.* 7: 241-263.
Unwin, E.E. 1932. On the structure of the respiratory organs of the terrestrial Isopoda. *Papers and Proceedings of the Royal Society of Tasmania* (1929-1932): 37-104.
Vandel, A. 1943. Essai sur l'origine, l'evolution et la classification des Oniscoidea (isopodes terrestres). *Bull. Biol. Fr. Belg.*_(Suppl.) 30: 1-136.
Vandel, A. 1960. Isopodes terrestres, premiere partie. *Faune Fr.* 64: 1-416.
Vandel, A. 1962. Isopodes terrestres deuxieme partie. *Faune Fr.* 66: 417-931.
Vandel, A. 1965. Sur l'existence d'oniscoides très primitifs menant une vie aquatique et sur le polyphylétisme des isopodes terrestres. *Ann. Spéléo.* 20: 489-518.
Van Name, W.G. 1936. The American land and freshwater isopod Cructacea. *Bulletin of the American Museum of Natural History* 77: 1-535.
Verhoeff, K.W. 1917. Zur Zenntnis der Atmung und der Atmungsorgane der Isopoda-Oniscoidea. *Biologisches Zentralblatt* 37: 113-127.
Verhoeff, K.W. 1920. Uber die atmung der Landasseln Zugleich ein Beitrag zur Kenntnis der Entstehung der Landteire. *Zeitschrift fur wissenschaften Zoologischen* 118: 365-447.
Warburg, M.R. 1968. Behavioural adaptations of terrestrial isopods. *Am. Zool.* 8: 545-559.
Warburg, M.R., K.E. Linsenmair & K. Berkovitz. 1984. The effect of climate on the distribution and abundance of isopods. *Symp. Zool. Soc. London* 53: 339-367.
Wieser, W. 1972a. O/N ratios of terrestrial isopods at two temperatures. *Comp. Biochem. & Physiol.* 43A: 859-868.
Wieser, W. 1972b. Oxygen consumption and ammonia excretion in *Ligia beaudiana* Milne-Edwards. *Comp. Biochem. Physiol.* 43 A: 869-876.
Wieser, W. 1972c. A glutaminase in the body wall of terrestrial isopods. *Nature* 239: 288-290.
Wieser, W. 1984. Ecophysiological adaptations of terrestrial isopods: a brief review. *Symp. Zool. Soc. London* 53: 247-262.
Wieser, W. & G. Schweizer. 1970. A re-examination of the excretion of nitrogen by terrestrial isopods. *J. Exp. Biol.* 52: 267-274.
Wieser, W. & G. Schweizer. 1972. Der Gehalt an Ammoniak und freien Aminosauren, sowie die Eigenschaften einer Glutaminase bei *Porcellio scaber* (Isopoda). *J. Comp. Physiol.* 81: 73-88.

Wieser, W., G. Schweizer & R. Hartenstein. 1969. Patterns in the release of gaseous ammonia by terrestrial isopods. *Oecologia* 3: 390-400.

Wilson, W.J. 1970. Osmoregulatory capabilities in isopods: *Ligia occidentalis* and *Ligia pallasii*. *Biol. Bull.* 138: 96-108.

Wood, C.M., R.G. Boutilier & D.J. Randall. 1986. The physiology of dehydration stress in the land crab *Cardisoma carnifex*: respiration, ionoregulation, acid-base balance and nitrogenous waste excretion. *J. Exp. Biol.* 150: 271-296.

Wright, J.C. & J. Machin. 1990. Water vapour absorption in terrestrial isopods. *J. Exp. Biol.* 154: 13-30.

Wright, J.C. & J. Machin. 1993a. Atmospheric water vapour absorption and the water budget of terrestrial isopods (Crustacea, Isopoda, Oniscidea). *Biol. Bull.* 184: 243-253.

Wright, J.C. & J. Machin. 1993b. Active water vapour absorption in isopods: evidence for pressure cycling as a supplement to the colligative uptake mechanism. *Physiol. Zool.* 66: 193-215.

Wright, J.C. & M.J. O'Donnell. 1992. Osmolality and electrolyte composition of pleon fluid in *Porcellio scaber* (Crustacea, Isopoda, Oniscidea): implications for water vapour absorption. *J. Exp. Biol.* 164: 189-203.

Wright, J.C. & M.J. O'Donnell. 1993. Total ammonia concentration and pH of haemolymph, pleon fluid and maxillary urine in *Porcellio scaber* (Isopoda, Oniscidea): relationships to ambient humidity and water vapour uptake. *J. Exp. Biol.* 176: 233-246.

Wright, J.C., C.W. Holliday & M.J. O'Donnell. 1993. Na^+/K^+ – ATPase activity in the pleopods and hindgut rectum of terrestrial isopods: Implications for colligative water vapour absorbtion. *J. Crust. Biol.* 14: 522-529.

Wright, J.C., M.J. O'Donnell & J. Reichert. 1994. Effects of ammonia loading on *Porcellio scaber*: Glutamine and glutamate synthesis, ammonia excretion and toxicity. *J. Exp. Biol.* 188: 143-157.

Wright, J.C., S. Caveny, M.J. O'Donnell & J. Reichert. 1994. Changes in tissue and hemolymph amino acid levels in response to ammonia loading in the terrestrial isopod *Porcellio scaber*. *Physiologist* 37: A66.

Ligia: A prototypal terrestrial isopod

Thomas H. Carefoot & Barbara E. Taylor
Department of Zoology, University of British Columbia, Vancouver, Canada

ABSTRACT

Isopod crustaceans exemplify the evolutionary transition of animals from aquatic to terrestrial habits, and genus *Ligia* exemplifies the prototypal land isopod. Morphological characteristics of ligiid pleopods and oöstegites, physiological characteristics of oxygen consumption, water uptake, and nutrition, and behavioural characteristics of swimming and orienting ability, all number among ligiid features which are transitional between ancestral marine and fully terrestrial forms. While such features are transitional in the sense of sea-to-land evolution, they nonetheless adapt ligiids to be successful in their semiterrestrial habit.

1 INTRODUCTION

Isopod crustaceans are exemplary of the evolutionary transition of animals from aquatic to terrestrial habits, and exist in all gradations from strictly marine to strictly terrestrial forms. Of an estimated 12,000 total species of isopods, some 4000 are in the terrestrial Suborder Oniscidea (Schultz, personal communication), and these have diversified into habitats of shore, savannah, marsh, forest and desert. Speculated origins of oniscideans (which include *Ligia* in the family Ligiidae) are the marine family Cirolanidae in the Suborder Flabellifera (Vandel 1943) and the marine Suborder Asellota (Schmalfuss 1989). Within the Suborder Oniscidea it is fairly well accepted that evolution proceeded from a ligiid-like ancestral stock through oniscid- and armadillidiid-like descendants (see Fig. 1). Genus *Ligia* exemplifies the semiterrestrial phase of the sea-to-land continuum. *Ligia*, commonly known as rock lice or sea slaters, inhabit supralittoral areas worldwide. Their habitat is a narrow vertical range of shoreline from the wave-splash region to a few metres above. Their behaviour is generally nocturnal, with emergence from protective crevices at dusk to forage on seaweeds in the upper intertidal region. At least 7 common species of *Ligia* inhabit similar shoreline areas on various continents. Their adult sizes range from 2-8 cm.

With regard to features of sea-to-land evolutionary transition embodied by *Ligia*, most noteworthy are the morphological, physiological, and behavioural traits which

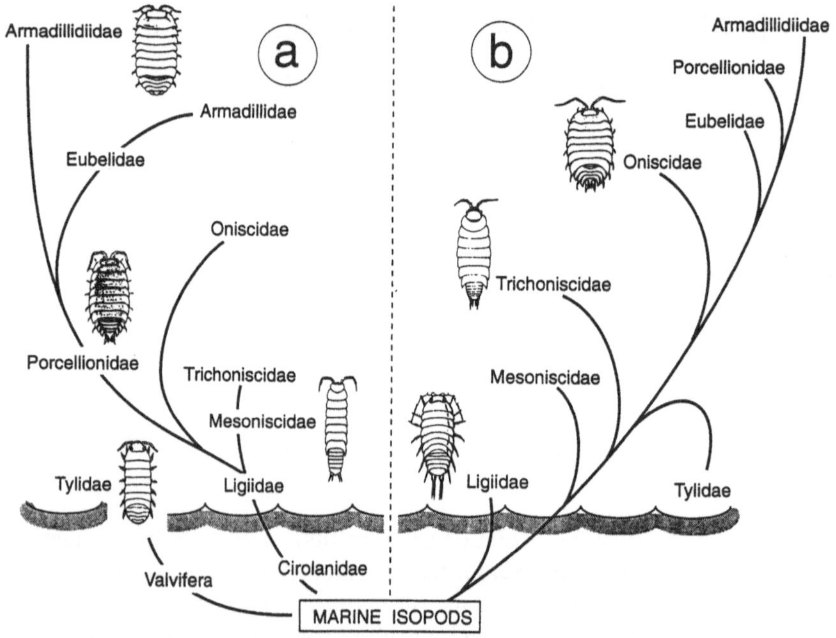

Figure 1. Proposed phylogenies of oniscidean isopods: a) Adapted from Vandel (1943); b) Adapted from Schmalfuss (1975, 1989). Isopod drawings from Sutton (1972) and Schmalfuss (1989).

arise when organisms exchange a seawater medium for an aerial one. As inhabitants of the border between land and sea, *Ligia* possess a combination of aquatic and terrestrial characteristics. Water, or the lack of it, has been fundamentally important in the evolution to terrestrial life. Uptake of water and water vapour, moistening of gas-exchange surfaces, irrigation of marsupia, and elimination of nitrogenous wastes are all issues of water relations which have featured prominently in the adaptation to terrestrial existence of ligiids and other oniscideans.

2 WATER

2.1 *Desiccation: A principal limiting factor*

Replenishment of water lost through desiccation in oniscideans is accomplished by direct uptake from standing water and from water-vapour absorption. Direct uptake can involve drinking or a unique system of water-conduction. The latter is present in all terrestrial isopods including *Ligia* (Verhoeff 1920; Hoese 1981, 1982a), and this system affords isopods some degree of resistance to desiccation. The order of increasing resistance to drying in four commonly studied general of oniscideans, namely, *Ligia* < *Oniscus* < *Porcellio* < *Armadillidium* (Miller 1938; Edney 1951; Bursell 1955; Cloudsley-Thompson 1956; Mayes & Holdich 1975; Carefoot et al. 1990) accords, as expected, with the order in which the groups are placed in evolution from sea to land (Miller 1938; Vandel 1943; Edney 1968), and this order also correlates with the nature of water-conducting systems possessed by these groups.

The most primitive system is possessed by *Ligia*. It consists of a series of interconnected shallow channels through which fluid moves by capillarity. Two main longitudinal ducts are sited external to the legs on the lateral-ventral body region. These link anteriorly to the openings of the bilateral maxillary glands (excretory organs) and terminate posteriorly in the region of the first pair of pleopods (gas-exchange organs). Urine from the maxillary glands is not released to the substratum in *Ligia* or other oniscideans; rather, the discharges flow in posterior direction via the water-conducting system to the pleopodal region where they moisten the gas-exchange surfaces. From there the urine flows to the anus where it may ultimately be absorbed in the rectum (Verhoeff 1920; Hoese 1981; pers. comm. Wright & Machin). Such rectal absorption has not been experimentally documented, but has been inferred from transport characteristics of the hindgut epithelium (deep, mitochondria-rich cells with apical and basal infoldings: Smith et al. 1969; Vernon et al. 1974), and the epithelium's known involvement in extracting water from the faeces (Kuenen 1959). As noted, ligiids can also pick up water from puddles and droplets, which joins the urinary flow in the channels. This is accomplished by dipping the closely opposed sixth and seventh walking legs as a unit into a droplet or puddle and allowing water to be conducted upwards via capillary action through the grooves formed between the legs (Hoese 1982a).

The *Ligia*-type conducting system differs from a more advanced *Porcellio*-type, found in the other more terrestrially evolved oniscideans. Here an additional dorsal component of ductules, comprised of the articulations between dorsal exoskeletal plates, allows fluid to flow along the soft cuticular sutures at each articulation from one lateral duct to the other (Hoese 1981). Hoese (1981) noted that the advanced *Porcellio*-type conducting system is a closed system through which only urine is conducted. The principal areas of moistening are the pleopodal gas-exchange surfaces and the marsupium in brooding females.

Not only do ligiids directly pick up water by uropodal dipping, but they also have the ability to absorb water vapour directly from the air. This has been confirmed for *Ligia oceanica*, as well as for 13 other species of terrestrial isopods by Wright & Machin (1990, 1993). The authors have shown that it is an active process functioning in net gain down to about 89% R.H. in most species. Wright & Machin (1990, 1993) have shown that WVA (= water vapour absorption) occurs specifically in the ventral abdominal or pleon region. This area, where the five pairs of gas-exchanging pleopods are attached, is normally bathed with a moistening fluid (Verhoeff 1920) isosmotic with the hemolymph (Wright & O'Donnell 1992). When animals are transferred from desiccating conditions to moist air, the isosmotic fluid is replaced with an hyperosmotic 'uptake' or pleon fluid, which has an osmolality as high as $8.2\ \mathrm{OsM} \cdot \mathrm{l}^{1}$, about 10-fold higher than that of the haemolymph (Wright & O'Donnell 1992, data for *Porcellio scaber*). As water vapour is absorbed by the fluid its osmolality declines with time to values close to determined equilibrium osmolality for ambient air.

Wright & Machin (personal communication) note that pleopodal beating enhances the uptake process. This metachronal ventilatory rhythm is present in all oniscideans. Not only does it facilitate ventilation of the branchial chamber and expose the hyperosmotic fluid directly to humid air, but it also increases pressure within the branchial chamber, thereby increasing the driving force for water uptake.

Ligia's transitional sea-land habitat with exposure to sea spray and access mainly to saltwater puddles and seaweeds as foods provides potential for severe salt-loading and concomitant osmoregulatory demands. In fact, *Ligia*'s haemolymph osmotic concentration is highest of all oniscideans, usually exceeding that of the seawater bathing the habitat (reviewed in Carefoot 1993). Such characteristics of *Ligia* prompted Wright & Machin (1993) to speculate that water-vapour absorption in littoral and supralittoral isopods might have evolved primarily to provide a salt-free water source for osmoregulatory purposes, rather than a source of water per se. In their view, secretion of a hyperosmotic pleon fluid in ancestral littoral forms would have been of selective value for ridding the body of excess salts gained through evaporative concentration of body fluids as well as dietary salt-loading. This, in turn, would have conferred an obvious preadaptation for water-vapour absorption and provided an evolutionary boost to land colonization by oniscideans.

2.2 *Ammonotely is the major excretory mode in Ligia and other terrestrial isopods*

Just as in marine forms, terrestrial isopods rely mainly on the release of ammonia to rid the body of nitrogenous waste. However, whereas for marine isopods this is in the form of ammonium ion (NH_4^+), for oniscideans it is in the form of ammonia gas (NH_3) (Dresel & Moyle 1950). Approximately 95% of total nitrogenous waste is eliminated via NH_3 volatilization (Hartenstein 1968; Wieser & Schweizer 1970). Thus, ligiids and other oniscideans are exceptions to Needham's Rule which generalizes that ureotely or uricotely accompany evolution to terrestrial life. Wieser (1972) has suggested that this excretory mode is an evolutionary compromise. Retention of ammoniotely via gaseous ammonia release allows oniscideans to conserve energy as it is a less costly means of excreting nitrogen, but avoids substantial water loss. Excretion of such a toxic nitrogenous compound may be facilitated by the comparatively low excretion of nitrogen by isopods in general (Dresel & Moyle 1950).

3 GAS EXCHANGE

3.1 *Ligia's gas-exchange surfaces are reminiscent of an aquatic ancestry*

The gas-exchange surfaces of *Ligia*'s pleopods are primitive. In *Ligia*, as in all isopod crustaceans, the branchial chamber is situated in the ventral abdominal or pleon area and consists of five pairs of front-to-back overlapping pleopods. Each pleopod is double, with an innermost (closest to body) endopod and outermost exopod. At rest the exopods fit snugly together and form a protective cover for the branchial chamber. The pleopods can beat in synchronous rhythm, thereby ventilating the branchial chamber to facilitate gas exchange, and enhancing water-vapour absorption and ammonia excretion (see Wright & O'Donnell, present volume). While the endopods are similarly structured, thick-walled, flattish blood-containing sacs in all genera (Unwin 1932), the exopods in advanced forms are variously modified with features which enhance gas exchange. Although some authors consider that the different modes of gas-exchange have evolved independently, it is convenient for our purposes to consider them as an evolutionary transition, from relatively undifferen-

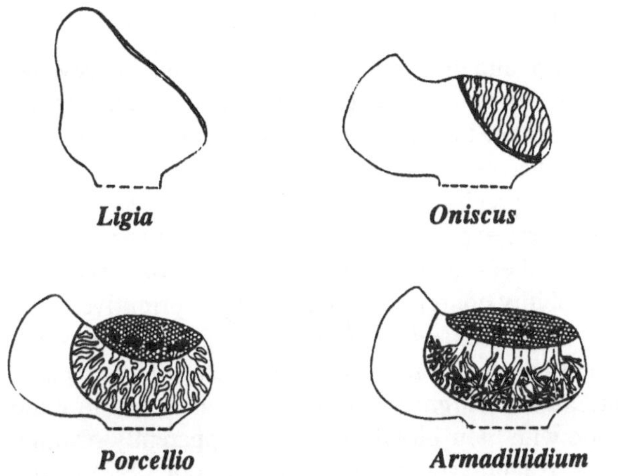

Figure 2. Pleopod (exopod) morphology for representative oniscideans (adapted from Ferrara et al. 1990).

tiated flattened sac-like structures in marine forms and in the semi-terrestrial *Ligia*, to ones with the exopodal borders extended to form corrugations or small lung-like cavities as in oniscids, to ones in which varying degrees of invaginations create 'pseudotracheae' as in porcellionids, armadillidiids, eubelids, and other advanced groups (Fig. 2; Hoese 1982b; Ferrara et al. 1990). Several authors have noted a correlation of gas-exchange morphology with habitat occupied in terrestrial isopods (Warburg 1968; Hoese 1982b; Ferrera et al. 1990), with unspecialized forms such as *Ligia* being restricted to hygric habitats, forms with lung-like pleopods being restricted to wet-forest biotypes, and those with pseudotracheate pleopods inhabiting arid biotypes.

The extent to which gas exchange occurs over the general body surface as compared with the pleopodal surfaces is also evolutionarily related. By blocking the pleopodal surfaces with emulsion paint, Edney & Spencer (1955) showed reductions in oxygen consumption of about 50% in *Ligia oceanica* and *Oniscus asellus*, 65% in *Porcellio scaber*, and 75% in *Armadillidium vulgare*, an order corresponding with one based on integument permeability.

3.2 *Drying of gas-exchange surfaces impairs oxygen consumption*

Related to general desiccation resistance is the ability of oniscideans to tolerate drying of their gas exchange surfaces. It comes as no surprise that comparatively severe desiccatory stress (as in dry air) leads to significantly reduced oxygen uptake ($\dot{V}O_2$) and ultimately to death in a genus-effect order of *Ligia* < *Oniscus* < *Porcellio* < *Armadillidium*. However, over short term, stress-effects of desiccating humidities might induce initially high $\dot{V}O_2$, with subsequent decline in $\dot{V}O_2$ as gas-exchange surfaces become increasingly drier. In a comparison of $\dot{V}O_2$ in four oniscidean species (Ligia pallasii, *Oniscus asellus*, *Porcellio scaber*, and *Armadillidium vulgare*), lowest oxygen uptake rate was in 100% R.H., presumably associated with the least stressful humidity, and highest for all species were in 76 and 55% R.H. treatments (Carefoot et al. 1990).

3.3 *An ancestral ability for underwater gas exchange is retained in ligiids*

Although *Ligia* is principally terrestrial and does not usually spend time in seawater, there are occasions when they enter this environment. For example, when foraging for intertidal seaweeds or during exposure to storm surge in their supralittoral habitat, *Ligia* undoubtedly are washed by waves into the water. *Ligia* is also seen to enter shallow tidepools voluntarily or, more precipitously, to tumble from rocks and cliff edges when in escape flight. Immersion experiments on several *Ligia* species have shown that they survive almost indefinitely in cool seawater (Tait 1916; Parry 1953, Taylor & Carefoot 1990, 1993), an ability conferred, in part, by their primitive undifferentiated pleopodal gas-exchange surfaces which function similarly to those in strictly marine forms. Species better adapted to terrestrial life, such as *Oniscus asellus, Porcellio scaber,* and *Armadillidium vulgare*, survive less than a day in either freshwater or in seawater isosmotic with their blood, with death apparently resulting from oxygen deprivation rather than osmotic stress (Taylor & Carefoot 1993). Figure 3 shows that $\dot{V}O_2$ of submerged *A. vulgare* declines to near-zero within 24h, while that of *Ligia pallasii* holds steady or even increases over a several-day period (Taylor & Carefoot 1993), reflecting the more primitive gas-exchange mechanisms of the latter. Figure 4 shows that underwater $\dot{V}O_2$ proficiency for a range of oniscideans with increasing gas-exchange specialization, measured within 1h of immersion, declines from *Ligia* (159% of $\dot{V}O_2$ in air), to *Oniscus* (44%), *Porcellio* (67%), and *Armadillidium* (52%). The data are interesting, not just for the reason that their order parallels closely the degree of terrestriality of each genus but, rather, for the fact that there is at least some retention in terrestrial isopods of an ancestral ability for underwater $\dot{V}O_2$. The higher aquatic respiratory rate of *Ligia* as compared with its aerial rate is at least partly explained by the fact that its pleopods beat vigorously and continuously when underwater (125-130 · min^{-1} for adult-size *Ligia* species at 15°C: Numanoi 1933; Ellenby 1951; Taylor & Carefoot 1993), a behaviour absent in other

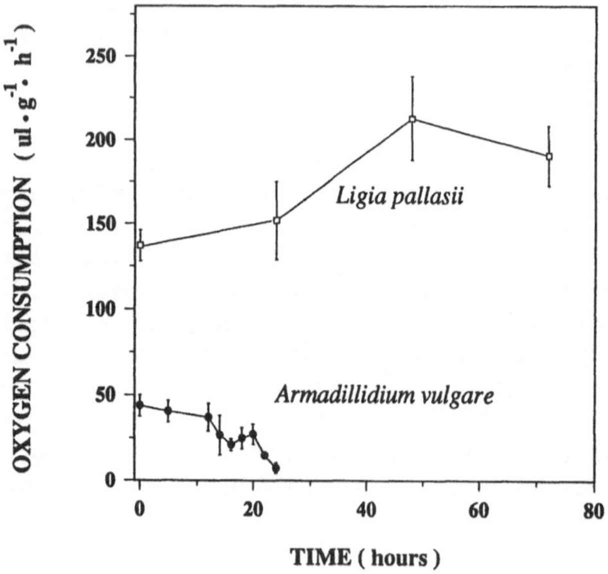

Figure 3. $\dot{V}O_2$ over time for *Ligia pallasii* and *Armadillidium vulgare* at 15°C. Each point represents the mean value (± standard error) of three animals (adapted from Taylor & Carefoot 1993).

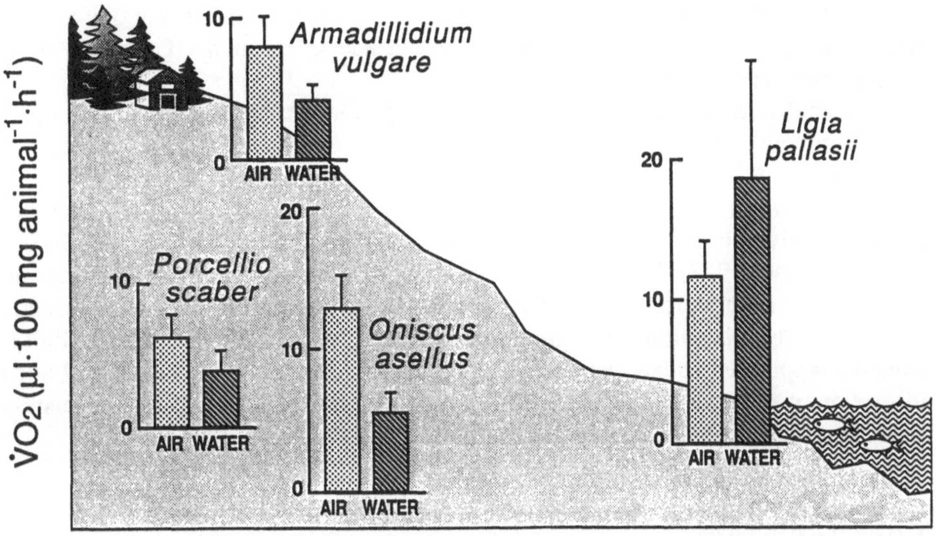

Figure 4. Aerial versus aquatic $\dot{V}O_2$ in terrestrial isopods (adapted from Taylor & Carefoot 1993).

oniscideans when submersed. Of course, pleopodal beating is common in marine forms, where it serves the dual purpose of ventilating the branchial area and propelling the animal.

3.4 *Aerial oxygen consumption by Ligia is relatively high*

There appear to be functional differences in gas-exchange proficiency in oniscideans in accordance with the degree of specialization of the pleopods. For example, in a comparative study of aerial oxygen consumption in *Ligia pallasii*, *Oniscus asellus*, *Porcellio scaber*, and *Armadillidium vulgare*, the oxygen uptake data may be separated into two statistically homogenous subgroups: one, a higher relative rate represented by *L. pallasii* and *O. asellus*; the other, a lower relative rate represented by *P. scaber* and *A. vulgare* (Taylor & Carefoot 1993). This is pertinent from the standpoint of pleopodal specialization since, as noted earlier, *Ligia* and *Oniscus* rely on pleopodal gas-exchange surfaces which are mainly undifferentiated, while the other two species possess pseudotracheate pleopods. Thus, a higher basal metabolic rate of the 'pleopodal breathers' in comparison with the 'pseudotracheal breathers' is indicated and accords with the general behaviour of these four species. *Ligia* and *Oniscus* are generally more active and faster moving than the other two genera. During collection when the animals are uncovered, for example, *Ligia* and *Oniscus* are more prone to run, while *Porcellio* initially lies motionless and *Armadillidium* curls into a ball.

4 NUTRITION

Nutritional needs of *Ligia* are fulfilled by an ancestral diet of seaweed. Marine isopods are primarily herbivorous, feeding on a variety of algal, diatom, and other plant growths. Reminiscent of this ancestry, *Ligia* species eat encrusting diatoms and macrophytes, exhibiting a preference for brown and green algae (Carefoot 1973a; Koop & Field 1980; Willows 1987).

In general little is known of nutritional needs for terrestrial isopods. However, specific mineral requirements are known for some oniscideans. Salts are required generally for osmotic regulation (Barnes 1940; Parry 1953; Todd 1963; Wilson 1970), calcium for growth (Numanoi 1934; Inagaki 1971; Carefoot 1984), and copper for hemolymph production and other metabolic processes (Dallinger 1977; Dallinger & Wieser 1977). Unlike their marine counterparts, terrestrial isopods are not bathed in a medium from which they can absorb these nutrients; hence, essential minerals must be acquired through consumption of food. In the transitional ligiids, uropodal dipping into seawater puddles and possibly drinking of seawater will allow absorption of minerals, reminiscent of the ancestral marine forms' direct acquisition of such nutrients from their seawater medium.

5 REPRODUCTION

5.1 *The ancestral marsupium is modified for terrestrial life*

The ancestral oniscidean marsupium presumably resembled one of three primary types found in extant marine isopods (Schultz 1969). The most common marine type is an open one in which the oöstegites extend from the base of each walking leg inwards to overlap loosely with those from the opposite side. A pouch is thus formed, in which fertilized eggs are deposited. Although five pairs of oöstegites are generally present in marine forms, numbers vary in some species from four to one (Schultz, 1969). There has been little modification of this basic plan in semiterrestrial ligiid descendants. Here, the broadly overlapping oöstegite plates meet imperfectly with the ventral body wall at front and back, leaving openings into the marsupium, smaller than those found in marine isopods but not completely sealed as in advanced oniscideans. By use of dyes, Hoese (1984) showed that in *Ligia,* water taken up from puddles via capillary action enters the rearward marsupial opening and moves anteriorly through the marsupium, providing the eggs and later embryonic stages with water and oxygen. Additionally, but not specifically noted by Hoese (1984), pleopodal dipping must also contribute to marsupial irrigation in *Ligia.* Provision of external water is the sole maternal contribution to *Ligia*'s 'amphibian'-type marsupium. All other terrestrial isopods have a closed or 'terrestrial'-type marsupium, which differs from the more primitive *Ligia* type in that it is completely sealed both front and back. Here, special maternal extensions into the marsupium, called cotyledons, supply the developing eggs and embryos with nutritional substances (Hoese & Janssen 1989). The secretion of a nutritional fluid via the cotyledons constitutes a level of maternal care which surpasses that of ligiids or littoral isopods, whose only maternal care is the provision of a moist environment. In both types, parturition involves the female flexing its abdomen upwards which causes the posterior pairs of oostegites to gape and allows egress of the youngsters.

5.2 *Reproductive patterns of Ligia are similar to other isopods*

Apart from the marsupial differences already noted, there have been no other obvious modifications in breeding patterns which would set *Ligia* apart from other oniscideans or, indeed, from their marine ancestors (e.g. cirolanids; Vandel 1943). The basic pattern for *Ligia* is outlined in Table 1 and this differs in only minor details from patterns for various marine, freshwater, and terrestrial forms (Willows 1983; Warburg 1987). Species of *Ligia,* as well as other oniscideans, are basically iteroparous, reproducing more than once in their lifetimes (Table 1; also Warburg 1987). This differs from the pattern in marine forms which tends more to semelparity (Warburg 1987). The 2-2.5 year life-span in *Ligia* species also appears to be somewhat longer than in marine forms but shorter than in other oniscideans (Warburg et al. 1984).

6 LOCOMOTION

6.1 *Underwater locomotory ability is superior in Ligia*

As part of our general study on the ability of *Ligia* to live underwater – a foreign, albeit ancestrally compatible environment – we assessed its swimming capability in comparison with other terrestrial isopods (Taylor & Carefoot 1990, 1993). Whereas ligiids seem to retain an ancestral ability to swim, none of several terrestrial oniscids, porcellionids, or armadillidiids tested, could do so. Also, ligiids are generally more active when immersed. When prevented from swimming by fatigue or confinement, they will actively explore their aquatic environment by walking. This behaviour is absent or greatly limited in other oniscideans which tend to sit motionless when underwater. Their superior locomotory capability provides a means for *Ligia* to regain the shore after entering the sea through dislodgement by waves or predatory attack. *Ligia* of all ages are capable of swimming. Their swimming action combines an alternating arching and flexing of the abdomen with a synchronous paddling stroke of the foremost four pairs of legs. Except for an up and down wobbling caused by the swimming movements, a fairly straight path is followed. The movement is reminiscent of a human butterfly stroke, and is unlike the swimming motion of most aquatic species which use their pleopods and travel in a straight line. The action in *Ligia* occurs at a frequency of about 3 complete sequences · sec^{-1} at 15°C (Taylor & Carefoot 1990).

Our experiments to test swimming proficiency of *Ligia exotica* led to the following conclusions: (Taylor & Carefoot 1990): 1) *Swimming velocity*: *Ligia* swims at 3-5 body lengths · sec^{-1}; 2) *Endurance*: results were variable but an average-length animal of 30 mm could swim for about 22 sec before tiring; 3) *Experience:* age or repeated daily swimming in experiments does not significantly improve swimming ability in *Ligia.*

Ligia's proficiency at swimming is associated with its superior ability to use oxygen from an aquatic medium. As noted earlier, *Ligia*'s aquatic metabolic rate is approximately twice that of *Oniscus, Porcellio,* and *Armadillidium* for equal-sized animals, although part of this enhanced rate is attributed to the active beating by *Ligia* of its pleopods when submerged.

Table 1. Life cycles and reproductive data for species of *Ligia*.

Species	Geographical area	Length of life (yrs)	Sex ratio (m : f)	Breeding season*	Minimum age at first breeding (mos)	No of broods per season	No of eggs per brood	Comments	Reference
dilatata	Cape Peninsula, South Africa	2	1 : 1.05	Spring-autumn	12	1–		Not possible to estimate how many females survive to breed a second time	Koop & Field 1980
oceanica	Yorkshire, United Kingdom	2.5	1 : 1.07	April-August	12	1–2	135–204 (early breeders) 93–118 (late breeders)	Only 10% of females survive to reproduce a second time in the season	Willows 1987 and Sutton et al. 1984
oceanica	Devon, United Kingdom	2.5		All year, spring peak	12	1–2	80 (40–100)	Based on lab studies	Nicholls 1931
pallasii	British Columbia, Canada	1.5–2	1 : 1	May-June	12	1–2	48 ± 11 (S.D.)		Carefoot 1973b

*Defined as females bearing eggs or juveniles in their marsupium.

6.2 *Underwater orientation and navigation by Ligia*

Investigations of swimming and navigational abilities in *Ligia* determined how these semiterrestrial isopods return to shore after entering the sea through dislodgement by waves or predatory attack (Taylor & Carefoot 1990). Under different circumstances *Ligia* might rely on different strategies to escape from the sea. For example, if an animal entered the water close to shore during daylight, its orienting behaviours and swimming ability would allow rapid return to land. If, however, it entered at night it could sustain itself by underwater gas exchange and feeding until it regained the shore by crawling upslope or until daylight returned to provide illumination and allow correct visual orientation.

In daylight situations the distinction, for *Ligia*, between what is close or far from shore is defined by its ability to orient and swim. Orientation experiments showed that direction-finding by *Ligia exotica* in water appeared to be governed chiefly by whether features of shoreline or bottom topography were visible and by degree of shading, while the sun was of secondary importance as a navigational cue (Taylor & Carefoot 1990). As indicated, our investigation of swimming showed that, on average, *Ligia* travel 3-5 body lengths · sec^{-1} for about 20 sec, regardless of their previous swimming experience. Therefore, an average-sized animal could swim 2 m before it became exhausted. This limitation is congruent with orienting ability, since we found *Ligia* cannot effectively orient to shore beyond a distance of about 3 m. Thus, if an animal was capable of swimming longer distances, it would be at heightened risk of swimming in the wrong direction.

Close proximity (*e.g.* 0.5-1.0 m) to shore provides visual navigational cues. At shallow depths, with the bottom sloping away at an acute angle, *Ligia* appears to differentiate between the lighter bottom features of the shallows and the darker-appearing depths, and swim towards shore. Where bottom features are less distinct, such as further offshore, swimming orientation becomes more random and position of the sun comes more into play. Where bottom features are absent, as in open-water, the sun becomes the principal navigational cue and swimming direction is counter to sun direction, even though the shore may be closer in the other direction. Photonegativity has been demonstrated for other ligiids (Perttunen 1961).

It is unknown if the moon serves as a similar phototactic cue for *Ligia* swimming in open water at night. However, it is doubtful that such behaviour has evolved. The only littoral isopod known to exhibit menotactic orientation is *Tylos latreilli* (Pardi 1954), and it has been suggested by Hoffmann (1989) that this adaptation is peculiar to this species as it is the only littoral isopod known to forage up to 50 m inland at night and to return to the littoral zone before dawn.

7 CONCLUSIONS

In summary, we have shown that *Ligia* is indeed a prototypal land isopod. There are numerous features of its morphology, physiology, and behaviour which are transitional between ancestral marine and fully terrestrial forms. Present-day ligiids do not represent the ancestral stock which gave rise to more terrestrial forms; rather, all these oniscideans are descendent from a common ligiid-like stock. *Ligia* is not a group whose adaptational failures stalled them at the semiterrestrial stage; rather, they are a group whose adaptational successes have enabled them to thrive in a complex habitat which straddles land and sea.

ACKNOWLEDGEMENTS

B.E.T. was supported during this work by a University Graduate Fellowship from the University of British Columbia, T.H.C. by a Research Grant from the Natural Sciences and Engineering Research Council of Canada.

REFERENCES

Barnes, T.C. 1932. Salt requirements and space orientation of the littoral isopod *Ligia* in Bermuda. *Biol. Bull.* 63: 496-504.

Bursell, E. 1955. The transpiration of terrestrial isopods. *J. Exp. Biol.* 32: 238-255.

Carefoot, T.H. 1973a. Feeding, food preference and uptake of food energy by the supralittoral isopod *Ligia pallasii. Mar. Biol.* 18: 228-236.

Carefoot, T.H. 1973b. Studies on the growth, reproduction, and life cycle of the supralittoral isopod *Ligia pallasii. Mar. Biol.* 18: 302-311.

Carefoot, T.H. 1984. Studies on the nutrition of the supralittoral isopod *Ligia pallasii* using chemically defined artificial diets: assessment of vitamin, carbohydrate, fatty acid, cholesterol and mineral requirements. *Comp. Biochem. Physiol.* 79A: 655-665.

Carefoot, T.H. 1993. Physiology of terrestrial isopods: a review. *Comp. Biochem. Physiol.* 106A: 404-422.

Carefoot, T., M. Mann, & S. Kalwa 1990. The effect of desiccation on oxygen uptake in terrestrial isopods. In P. Juchault & M. Moquard (eds.) *Biology of Terrestrial Isopods III, Proc Third International Symposium, Poitiers, July 10-12, 1990*, pp. 157-164.

Cloudsley-Thompson, J.L. 1956. Studies in diurnal rhythms VII. Humidity responses and nocturnal activity in woodlice (Isopoda). *J. Exp. Biol.* 33: 576-582.

Dallinger, R. 1977. The flow of copper through a terrestrial food chain III. Selection of an optimum copper diet by isopods. *Oecologia (Berl.)* 30: 273-276.

Dallinger, R. & W. Wieser 1977. The flow of copper through a terrestrial food chain I. Copper and nutrition in isopods. *Oecologia (Berl.)* 30: 253-264.

Dresel, E.I.B. & V. Moyle 1950. Nitrogenous excretion of amphipods and isopods. *J. Exp. Biol.* 27: 210-225.

Edney, E.B. 1951. The evaporation of water from woodlice and the millipede *Glomeris. J. Exp. Biol.* 28: 91-115.

Edney, E.B. 1968. Transition from water to land in isopod crustaceans. *Amer. Zool.* 8: 309-326.

Edney, E.B. & J.O. Spencer. 1955. Cutaneous respiration in woodlice. *J. Exp. Biol.* 32: 256-269.

Ellenby, C. 1951. Body size in relation to oxygen consumption and pleopod beat in *Ligia oceanica (L.) J. Exp. Biol.* 28: 492-507.

Ferrera, F., P. Paoli, & S. Taiti 1990. Morphology of the pleopodal lungs in the Eubelidae (Crustacea, Oniscidea). In P. Juchault & M. Moquard (eds.) *The Biology of Terrestrial Isopods III, Proc. Third International Symposium, Poitiers July 10-12, 1990*, pp. 9-16.

Hartenstein, R. 1968. Nitrogen metabolism in the terrestrial isopod, *Oniscus asellus. Amer. Zool.* 8: 507-519.

Hoese, B. 1981. Morphologie und Funktion des Wasserleitungssystems der terrestrischen Isopoden (Crustacea, Isopoda, Oniscoidea). *Zoomorph.* 98: 135-167.

Hoese, V.B. 1982a. Der *Ligia*-typ des Wasserleitungs systems bei terrestrischen Isopods und seine Entwicklung in der Familie Ligiidae (Crustacea, Isopoda, Oniscoidea). *Zool. Jb. Anat.* 108: 225-261.

Hoese, B. 1982b. Morphologie und Evolution der Lungen bei den terrestrichen Isopoden (Crustacea, Isopoda, Oniscoidea). *Zool. Jb. Anat.* 107: 396-422.

Hoese, B. 1984. The marsupium in terrestrial isopods. *Symp. Zool. Soc. Lond.* 53: 65-76.

Hoese, B. & H.H. Janssen 1989. Morphological and physiological studies on the marsupium in terrestrial isopods. *Monitore Zool. Ital. (N.S.) Monogr.* 4: 153-173.

Hoffmann, G. 1989. The orientation of terrestrial isopods. *Monitore Zool. Ital. (N.S.) Monogr.* 4: 489-512.

Inagaki, H. 1971. Variabilité du poids et de la teneur en calcium de l'exuvie du Crustacé Isopode *Ligia oceanica* (L.) en rapport avec sa sénescence. *Comp. Ren. Séanc. Soc. Biol.* 165: 5-9.

Koop, K. & J.G. Field 1980. The influence of food availability on population dynamics of a supralittoral isopod, *Ligia dilatata* Brandt. *J. Exp. Mar. Biol. Ecol.* 48: 61-72.

Kuenen, D.J. 1959. Excretion and waterbalance in some land-isopods. *Ent. Exp. Appl.* 2: 287-294.

Mayes, K.R. & D.M. Holdich 1975. Water exchange between woodlice and moist environments with particular reference to *Oniscus asellus. Comp. Biochem. Physiol.* 51A: 295-300.

Miller, M.A. 1938. Comparative ecological studies on the terrestrial isopod Crustacea of the San Francisco Bay region. *Univ. Calif. Publ. Zool.* 43: 113-141.

Numanoi, H. 1933. Effect of temperature on the frequency of the rhythmical beatings of the pleopods of *Ligia exotica. J. Fac. Sci. Tokyo Imp. Univ.* 3: 217

Numanoi, H. 1934. Calcium in the blood of *Ligia exotica* during non-molting and molting phases. J. Fac. Sci. Tokyo Imper. Univ. 3: 351-358.

Nicholls, A.G. 1931. Studies on *Ligia oceanica.* I. A. Habitat and effect of change of environment on respiration. B. Observations on moulting and breeding. *J. Mar. Biol. Assoc. U.K.* 17: 655-673.

Pardi, L. 1954. Über die Orientierung von *Tylos latreillii* Aud. & Sav. (Isopoda terrestria). *Zeitschr. Tierpsychol.* 11: 175-181.

Parry, G. 1953. Osmotic and ionic regulation in the isopod crustacean *Ligia oceanica. J. Exp. Biol.* 30: 567-574.

Perttunen, V. 1961. Réactions de *Ligia italica (F.)* a la lumière et l'humidité de l'air. *Vie et Milieu* 12: 219-259.

Price, J.B. & D.M. Holdich 1980. An ultrastructural study of the integument during the moult cycle of the woodlouse, *Oniscus asellus* (Crustacea, Isopoda). *Zoomorph.* 95: 250-263.

Schmalfuss, H. 1975. Morphologie, Funktion, und Evolution der Tergithöcker bei Landisopoden (Oniscoidea, Isopoda, Crustacea). *Z. Morph. Tiere.* 80: 287-316.

Schmalfuss, H. 1989. Phylogenetics in Oniscidea. *Monitore Zool. Ital. (N.S.) Monogr.* 4: 3-27.

Schultz, G.A. 1969. *The Marine Isopod Crustaceans.* Dubuque, Iowa: Wm. C. Brown Co.

Smith, W.J., E.R. Witkus, & R.S. Grillo. 1969. Structural adaptations for ion and water transport in the hindgut of the wood louse *Oniscus asellus. J. Cell Biol.* 43: 135a-136a.

Sutton, S. 1972. *Woodlice.* London: Ginn & Co.

Sutton, S.L., M. Hassall, R. Willows, R.C. Davis, A. Grundy, & K.D. Sunderland 1984. Life histories of terrestrial isopods: a study of intra- and interspecific variation. *Symp. Zool. Soc. Lond.* 53: 269-294.

Tait, J. 1916. IV.- Experiments and observations on Crustacea: Part I. Immersion experiments on *Ligia. Proc. Roy. Soc. Edinb.* 37: 50-58.

Taylor, B. & T. Carefoot. 1990. Underwater survival, orientation, and locomotory ability in the semiterrestrial Japanese 'Funamushi' *Ligia exotica* (Roux). In P. Juchault & M. Moquard (eds.) *Biology of Terrestrial Isopods III, Proc. Third International Symposium, Poitiers, July 10-12, 1990,* pp. 121-139.

Taylor, B.E. & T.H. Carefoot 1993. Terrestrial life in isopods: evolutionary loss of gas-exchange and survival capability in water. *Can. J. Zool.* 71: 1372-1378.

Todd, M.E. 1963. Osmoregulation in *Ligia oceanica* and *Idotea granulosa. J. Exp. Biol.* 40: 381-392.

Unwin, E.E. 1932. On the structure of the respiratory organs of the terrestrial Isopoda. *Papers & Proc. Roy. Soc. Tasmania,* 1932: 37-104.

Vandel, A. 1943. L'origine, l'evolution et la classification des *Oniscoidea* (isopodes terrestres). *Bull. Biol. France Belgique Suppl.* 30:1-136.
Verhoeff, K.W. 1920. Über die atmumg der Landasseln, zugleich ein Beitrag zur Kenntnis der Entstehung der Landtiere. *Z. Wiss. Zool.* 118: 365-447.
Vernon, G.M., L. Herold, & E.R. Witkus 1974. Fine structure of the digestive tract epithelium in the terrestrial isopod, *Armadillidium vulgare. J. Morph.* 144: 337-360.
Warburg, M.R. 1968. Behavioral adaptations of terrestrial isopods. *Amer. Zool.* 8: 545-559.
Warburg, M.R. 1987. Isopods and their terrestrial environment. *Adv. Ecol. Res.* 17: 187-242.
Warburg, M.R., K.E. Linsenmair, & K. Bercovitz 1984. The effect of climate on the distribution and abundance of isopods. *Symp. Zool. Soc. Lond.* 53: 339-367.
Wieser, V.W. 1972. Oxygen consumption and ammonia excretion in *Ligia beaudiana* M.-E. *Comp. Biochem. Physiol.* 43A: 869-876.
Wieser, W. & G. Schweizer 1970. A re-examination of the excretion of nitrogen by terrestrial isopods. *J. Exp. Biol.* 52: 267-274.
Willows, R. 1984. Breeding phenology of woodlice and oostegite development in *Ligia oceanica* (L.) (Crustacea). *Symp. Zool. Soc. Lond.* 53: 469-485.
Willows, R.I. 1987. Population dynamics and life history of two contrasting populations of *Ligia oceanica* (Crustacea: Oniscidea) in the rocky supralittoral. *J. Anim. Ecol.* 56: 315-330.
Wilson, W. 1970. Osmoregulatory capabilities in isopods: *Ligia occidentalis* and *Ligia pallasii. Biol. Bull.* 138: 96-108.
Wright, J.C. & J. Machin 1990. Water vapour absorption in terrestrial isopods. *J. Exp. Biol.* 154: 13-30.
Wright, J.C. & J. Machin 1993. Atmospheric water absorption and the water budget of terrestrial isopods (Crustacea, Isopoda, Oniscidea). *Biol. Bull.* 184: 243-253.
Wright, J.C. & M.J. O'Donnell 1992. Osmolality and electrolyte composition of pleon fluid in *Porcellio scaber* (Crustacea, Isopoda, Oniscidea): implications for water vapour absorption. *J. Exp. Biol.* 164: 189-203.

Mechanism and function of turn alternation in *Armadillidium vulgare* (Latreille)

Tang Bo-Ping
Department of Biology, Yancheng Normal College, Yancheng, People's Republic of China
Gui Hong & Wu Tian
Department of Biology, Nanjing Normal University, Nanjing, Peoples Republic of China

ABSTRACT

In a series of experiments, under different conditions, with 3,963 woodlice, it was observed that 1) The turn alternation is manifested by the 1st and the 2nd leg on the side opposite to the first turn. The action is not related to the motion of other legs, 2) Isopods manifest an apparent behaviour of turn alternation, 3) The behaviour disappears with gradual increase in the distance between the forced turn and the free turn, 4) The turn alternation behaviour neither intensifies nor weakens with the increase of 'accurate' or 'reverse' forced turn. The behaviour is neither based on 'short-term memory' nor on the sensory input by the antennae.

1 INTRODUCTION

The turn-alternation behaviour in the woodlouse (*Armadillidium vulgare* Latreille) was first discovered by Kupfermann (1966). Since then, Hughes (1966, 1967, 1985, 1989) has detailed out the behaviour in *Porcellio scaber* Latreille. The present report describes the mechanism and the function of the turn-alteration behaviour, under variable experimental conditions, in *Armadillidium vulgare* Latreille.

2 MATERIAL AND METHODS

A. vulgare, used in various experiments, were collected from flower-beds, road-side ditches and under haystacks, and kept in laboratory in metal containers, containing damp soil, leaf litter and sliced potato at the bottom to provide food and moisture. The maze alleys (20mm wide and 20 mm deep) for the study were constructed of 5 mm thick transparent glass blocks. All maze alleys were constructed on a back table top and illuminated with a 1 m overhead 40 watt room fluorescent lamp. All experiments were performed in a darkened room, and each woodlice were tested only once.

3 RESULTS

3.1 *Free turn and turn alternation*

161 woodlice were individually placed in the starting alley in maze I (Fig. 1). 78 (48.4%) isopods reached point A, and 83 (51.6%) reached point B. In both case the animals did not show any tendency to either turn left or right ($X^2 = 0.04960$). Each of the 229 woodlice were forced to make a right or left turn in maze II (Fig.1) and then allowed a free choice at point A. Out of these, 209 (91.3%) reached point A_1 and 20 (8.7%) reached point A_2 ($X^2 = 154.3$, $P < 0.001$).

3.2 *Effect of varied distance between a forced turn and a free point*

A total number of 1,305 woodlice were forced through either a left or a right turn in maze II (Fig.1) and then allowed a free choice at any of the nine points proceeding the forced turn. The results are tabulated in Table 1. With the distance between the point of the forced turn and that of the free turn (D) as cross axis and the percentage

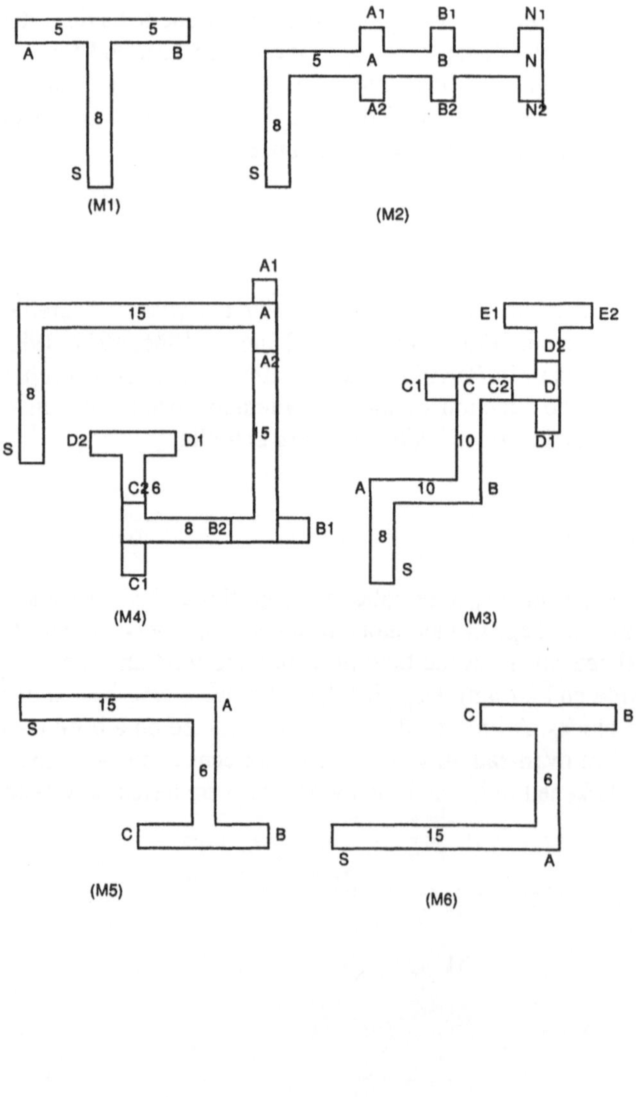

Figure 1. A diagrammatic illustration of turn alternation mazes (s = starting point; M = maze; figures re-present length of the maze in cm).

Table 1. Effect of a forced turn (D in cm) by adult *Armadillidium vulgare* on turn alternation at nine different distances from the forced turn point.

Distance between forced turn and free turn point (D)	Number of woodlice showing Turn alternation (n_1)	No turn alternation (n_2)	$p=n_1/(n_1+n_2)$ (%)
5	70	6	92.4
10.5	90	14	86.5
13	110	18	85.9
17.5	170	6	73.3
30.5	69	29	70.4
40	108	5	65.5
46.5	106	53	66.7
50	116	67	63.3
56	70	81	46.35

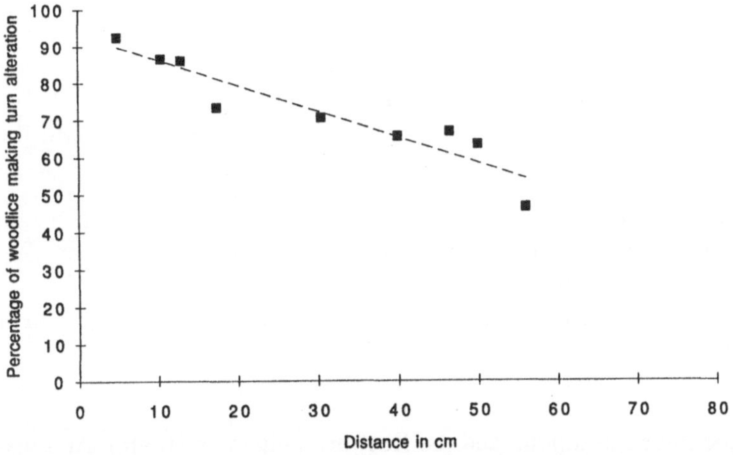

Figure 2. Percentage of adult *Armadillidium vulgare* showing turn alternation behaviour at nine distances from the forced turn.

of woodlice manifesting the turn behaviour as longitudinal axis, the plot is presented in figure 2. The correlation was calculated to be

$$P = 93.3 - 0.7D$$

correlation coefficient $r = -0.94$.

The relationship between D and P was significant ($t = r/Sr = -7.2$, $|t| > t_{0.01}$) at $P < 0.01$. These observations are somewhat different from those reported earlier by Kupfermann (1966), but agree with those made by Hughes (1967) with *Porcellio scaber*.

3.3 *Effect of detaining woodlice after a forced turn*

Alternation responses of 249 wood lice were observed after a forced turn in maze II. The distance between the starting point and the forced turn point was 8 mm. The

Table 2. Effect of detention after a forced turn (D in cm) on turn alternation behaviour in adult *Armadillidium vulgare.*

Detention time (D)	Number of woodlice showing Turn alternation (n_1)	No turn alternation (n_2)	$p = n_1/(n_1 + n_2)$ (%)
15	73	9	89.02
30	47	14	77.05
60	41	12	77.36
120	35	18	66.03

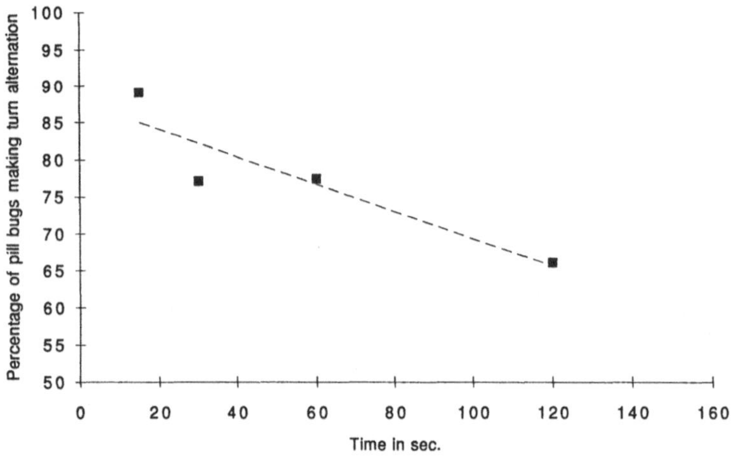

Figure 3. The effect of detention period following a forced turn on the turn alternation behaviour in adult *Armadillidium vulgare.*

distance between the forced turn point and the free turn point was 10 mm. At a distance of 5 mm from the forced turn point, the woodlice were detained by pressing their dorsum with a Chinese writing brush for the desired period. Isopods which could not be detained successfully, or those which did not resume walking immediately following the removal of the brush, were discarded. The results, summarized in Table 2, indicate that the percentage of isopods showing the turn alternation behaviour decreased with an increase in the detention time. The relationship was calculated to be

$$P = 87.76 - 0.185T$$
$$r = -0.914, \quad t = r/Sr = -3.186, \quad P < 0.01$$

The relationship between P and T was significant (Fig.3). The difference among the four groups was significant ($X^2 = 10.44$) at $P < 0.02$.

3.4 *Effect of 'correct' forced turn on turn alternation*

Each of 408 woodlice were placed in the starting alley of maze III and then forced into a 'correct' alternation turn once, twice or three times. The results are shown in Figure 4. The linear regression of P on N is not significant, and it shows that turn alternation is not affected by the frequency of 'correct' forced turns.

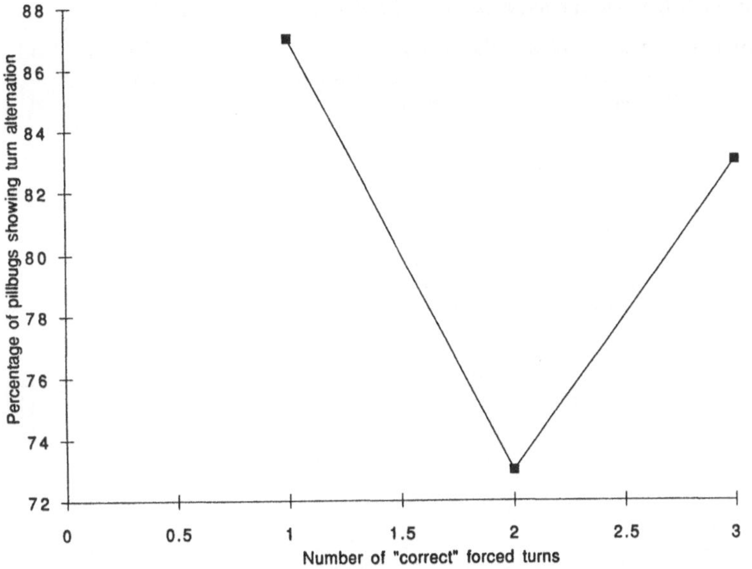

Figure 4. Effect of the number of 'correct' forced turns on the turn alternation behaviour in adult *Armadillidium vulgare.*

Table 3. Effect of 'contradictory' forced turns on the turn alternation behaviour in adult *Armadillidium vulgare.*

Number of 'contradictory' forced turns	Number of woodlice making Correcting turn (n_1)	No correcting turn (n_2)	$P = n_1/(n_1 + n_2)$ (%)
1	105	15	87.5
2	106	11	90.5
3	113	18	86.5

3.5 *Effect of 'contradictory' forced turn on turn alternation*

The effect of one to three 'contradictory' forced turns were studied in 338 woodlice in maze IV. The results can be tabulated in Table 3. The difference between the three groups of animals was not significant ($X^2 = 1.15$) at $P > 0.05$. The relationship of P with N was not clear. The turn alternation behaviour was not affected by the number of 'contradictory' forced turns.

3.6 *Turn alternation behaviour in mancas*

775 one year old mancas (total body length = 9mm) were allowed to walk in maze I. Results of the experiment are summarized in Table 4.

The relationship of D with P can be expressed as

$$P = 93.56 - 0.786D$$
$$r = -0.96, \quad t = r/Sr = 6.875.$$

The relationship of P with D is significant at $P < 0.01$ (Fig. 5).

Table 4. Turn alternation behaviour in mancas of *Armadillidium vulgare* (D in cm).

Distance between forced turn and free point (D)	Number of woodlice making Correcting turn(n_1)	No correcting (%) turn(n_2)	$P = n_1/(n_1 + n_2)$ (%)
20	81	22	78.64
30	91	42	68.42
40	121	84	59.02
45	42	25	62.68
50	70	54	56.45
55	63	70	47.37

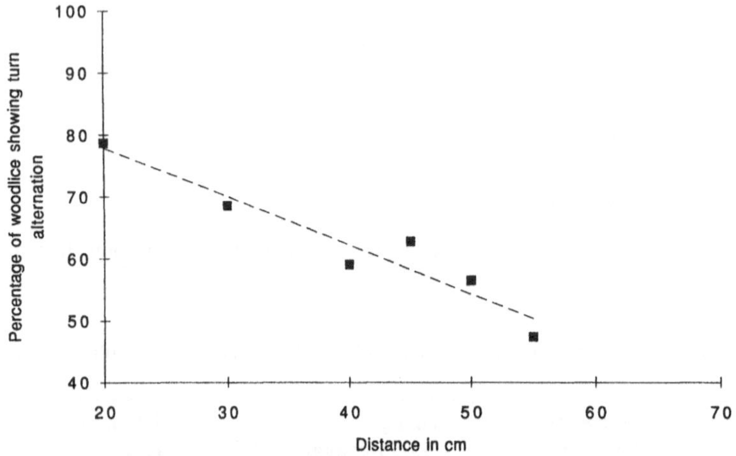

Figure 5. Percentage of mancas of Armadillidium vulgare showing turn alternation behaviour at nine distances from the forced turn.

The relationship of P with D in this experiment ($P = 93.56 - 0.786D$) is somewhat similar to that of experiment described in Section 3.2 ($P = 93.30 - 0.7D$). These regression coefficients in the two experiments were 0.786 and 0.7, respectively ($t = b_1 - b_2/Sb_1 - b_2 = 2.92$, $P > 0.01$). Thus, the difference between the results of the two experiments was not significant. This suggests that the behaviour of turn alternation is a innate behaviour, and is not an acquired behaviour.

3.7 *Effect of amputation of antennae on turn alternation behaviour*

54 woodlice with amputated antennae were provided with the turn alternation choice in maze II. Out of these, 48 woodlice turned in the opposite direction to a preceding turn at the free point. This results are significant ($X^2 = 31.3$) at $P < 0.01$.

3.8 *Effect of amputation of 1st and 2nd right legs on turn alternation*

102 woodlice with amputated 1st and 2nd right legs were allowed passage through maze V. Out of these, 93 (91.2%) animals turned in the alternate direction at the free-point. The percentage of woodlice making the turn was greater than chance at

$P < 0.001$. 103 woodlice with amputated 1st and 2nd right legs were force to turn left and then allowed to walk to a turn free point in maze VI. 53 of these woodlice reached point B while 50 of them did not. The difference between the two groups was not significant.

3.9 *Effect of amputation of 6th and 7th right legs on turn alternation behaviour*

100 woodlice with amputated 6th and 7th right legs were forced to turn right and then allowed to walk to a turn free point. 96 of these woodlice reached point B.
A second group of 100 woodlice with amputated 6th and 7th right legs were forced to turn left and then allowed to walk to a turn free point in maze VI. 88 of these woodlice reached point B. In each of the two above experiments the percentage of woodlice reaching point B was greater than chance at $P < 0.001$ level. However, difference between the two groups was not significant ($X^2 = 3.32$) at $P > 0.05$.

3.10 *Effect of amputation of 3rd, 4th and 5th right legs on turn alternation*

Turn alternation responses in 101 woodlice, with amputated 3rd, 4th and 5th right legs, which had been made to take a right turn, was observed in maze V. 92 (91.1%) of these woodlice turned in the opposite direction to the forced turns. A second group of 104 woodlice, with amputated 3rd, 4th and 5th right legs, were allowed to make a free turn after having been forced to take left turn. The percentage of woodlice showing the turn alternation behaviour was 87.5. The difference between the above two groups was not significant.

4 DISCUSSION

The results of the present study suggest that woodlice manifest turn alternation behaviour, and that turn alternation behaviour is prevalent in both mancas and adults. However, the difference between the group manifesting the turn alternation behaviour and the group that did not show this behaviour was not significant. It, therefore, appear likely that the behaviour is an innate behaviour and not an acquired characteristic.

In the experiments where woodlice were detained for 120 seconds following a forced turn, less than 50% of the animals ($P = 87.76 - 0.185T$) showed the turn alternation behaviour. Nevertheless, a majority of the woodlice which had been made to make a forced turn, took only about 45-60 seconds to cover a distance of 56 cm from the point of forced turn to the point of free turn. This implies that contrary to the findings of Kupfermann (1966), turn alternation is not mediated by a 'short-term' memory of the event. These conclusions are also supported by the results of the experiments described in Sections 3.4 and 3.5.

The observations upon animals with amputated legs suggest that the success of sequential turn alternation in wood lice may largely depend on the lateral thrusting influence of the 1st and the 2nd legs. The turn alternation behaviour is quite obvious in the animals with amputated 1st and 2nd legs on the same side of the direction of the forced turn. However, in animals with legs missing on the opposite side to the turn, the turn alternation behaviour is not obvious. Hughes (1989) obtained somewhat contradictory results with *Porcellio scaber*. Amputation of the last two (6th and 7th) legs or those of 3rd, 4th and 5th legs, in the present study, did not produce any obvious effect on turn alternation. Hughes (1989), on the other hand, insisted that the removal of last two legs in *Porcellio scaber* effects the turn alternation behaviour significantly. Removal of antennae has also been shown to have a significant effect of this behaviour.

The function of turn alternation in woodlice may be a way to ensure the quickest movement across an area containing different kinds of obstacles through a series of random responses in the same direction as the preceding turn. This would ensure that the animal 'escapes' quickly from a source of noxious stimulation, and finds with ease the food and the other essential elements for life. This behaviour may also keep the animal aware of the environment it is moving through at that time.

REFERENCES

Hughes, R.N. 1966. Some observation of correcting behaviour in wood lice (*Porcellio scaber*). *Animal Behaviour* 14:319.

Hughes, R.N. 1967. Turn alternation in wood lice (*Porcellio scaber*). *Animal Behaviour* 15:282-286.

Hughes, R.N. 1985. Mechanisms for turn alternation in wood lice (*Porcellio scaber*): The role of bilaterally asymmetrical leg movements. *Animal Learning & Behaviour* 13:253-260.

Hughes, R.N. 1989. Essential involvement of specific legs in turn alternation of the wood lice, *Porcellio scaber. Comp. Biochem. Physiol.* 93A (2): 493-497.

Kupfermann, I. 1966. Turn alternation in the pill bug (*Armadillidium vulgare*). *Animal Behaviour* 14:68-72.

Reproduction in woodlice: Flexibility to maximise individual fitness

J.M. Dangerfield
Department of Biology, University of Botswana, Gaborone, Botswana

S.R. Telford
Department of Zoology, University of Pretoria, Pretoria, South Africa

ABSTRACT

As fitness is a relative term, it is difficult to measure in nature. At present, longitudinal population studies and observations of resource allocation to different reproductive traits are the best sources of information for the development of theories to explain the evolution of life history tactics. Studies on terrestrial isopods suggest that tactics to maximise fitness are many and vary extensively even at the level of individual. Such flexibility, particularly in reproductive tactics, may be the key to the success of woodlice as terrestrial animals. In this review, we consider the available data on woodlice life histories from the evolutionary perspective of maintaining individual fitness. We consider the idea that proximate mechanisms must exist to allow the resource allocation 'decisions' within the overall reproductive pattern, and we discuss the constraints and limits to reproductive flexibility. These ideas are gathered into a suite of observations and testable hypotheses that we see as important for quantitative tests of reproductive flexibility as a strategy to maximise fitness in woodlice.

1 INTRODUCTION

Evolutionary biologists define fitness as the relative genetic contribution made by an individual to future generations, which, in real terms, translates into the success of an individual or phenotype in the production of offspring. Such a definition makes fitness difficult to quantify in real populations because it is a relative quantity; a function of the reproductive success of both current (present generation) and future (succeeding generations) competitors. Absolute or practical measures of individual fitness, such as the number of surviving offspring or lifetime reproductive success are difficult to measure in nature, consequently, indices of fitness such as the intrinsic rate of population increase, r, which is a population parameter, have become surrogates for individual fitness. Stearns (1976) has already noted that although intrinsic to the individual, life history traits that determine individual fitness are most frequently measured on populations.

A further difficulty is that numerical estimates of fitness cannot readily be compared between generations because of temporal and spatial variability in environments (Caswell 1983). Where resources and conditions vary in an unpredictable manner on time scales shorter than the generation time of the organism, fitness of the parental phenotype may be very different to that of offspring (Caswell 1989). For example, the production of 20 progeny may represent high fitness in the parental generation but a poor performance if repeated by offspring in the following generation.

Despite these constraints several attempts have been made to explain patterns in the various tactics that organisms display to maximise fitness, under the general heading of life history theory. Three broad groups of models have been developed: 1) Those that use optimality arguments to explain patterns in life history tactics that maximise fitness through a focus on how resources and environmental conditions determine specific aspects of population biology (Sibly & Calow 1986; Houston & McNamara 1992), 2) Models that consider general habitat components as indices of fitness conditions, notably the degree of stress (Southwood 1977), and 3) Models based on specific resource allocation trade-offs, for example the size and number of young (Smith & Fretwell 1974; Lloyd 1987). Habitat based models concentrate on the effects of factors extrinsic to the organism whilst resource allocation models focus on the intrinsic processes that result in life history trade-offs. Optimality approaches attempt to predict outcomes of the intrinsic trade-offs that will maximise fitness in response to the prevailing environmental conditions. There are very few critical tests of these models.

Detailed longitudinal studies of the population biology and life histories of woodlice have provided an extensive empirical base for the analysis of life history models (Sutton et al. 1984; Davis & Hassall in preparation). However, in this short review it is not our intention to consider the efficacy of life history models to explain variation in woodlouse life histories, but rather to consider the extent of the variability in reproductive traits from the evolutionary perspective of the individual, that is, the extent of flexibility available to allow 'decisions' on reproductive tactics capable of influencing individual fitness. The emphasis being on variances in key life history traits (e.g. Yoshimura & Clark 1991) rather than means and the sources of such variance.

Our objective is to take this theme and through an inductive process place the available information on woodlouse life history variation into an evolutionary framework. We then attempt to develop some of these explanations into future research objectives which include further observations and testable hypotheses.

2 WOODLICE POPULATION BIOLOGY AND LIFE HISTORIES

Woodlice are one of the few groups of organisms where significant long term field data have provided an understanding of the role of abiotic (McQueen & Carnio 1973; Warburg, Linsenmair & Bercovitz 1984; Warburg 1987; Hassall & Dangerfield submitted) and biotic (Paris & Pitelka 1962; Rushton & Hassall 1987; Hassall & Dangerfield 1990) factors in population dynamics. The necessity to obtain data on natality to complete analyses of these population trends has provided much information on the breeding biology (Lawlor 1976a; Sutton et al. 1984; Willows 1987) and reproductive patterns (Willows 1984; Souty-Grosset et al. 1988; Dangerfield & Hassall 1992) of several species. In addition there are a number of studies, notably on the temperate pillbug *Armadillidium vulgare* (Latr.), that have revealed mediators of the proximate mechanisms that determine these patterns (e.g. Mocquard, Juchault & Souty-Grosset, 1989). Collectively these studies suggest that woodlice employ a range of flexible reproductive tactics in response to environmental conditions.

2.1 *Breeding phenology*

Most temperate species are strict seasonal breeders, a response to favourable conditions for rapid brood development and offspring release (Willows 1984). Tropical species such as *Burmoniscus ocellatus* (Verhoeff), *Formosillo rafaelei* (Arcangeli) and *Orodillo maculatus* Arcangeli (Ma, Lam & Dudgeon 1991) and sub-tropical species such as *Bethalus pretoriensis* Dollfus (Dangerfield, unpublished data) also show seasonal reproduction, although breeding seasons are much longer than for most temperate species. There are some exceptions to this pattern such *as Porcellionides pruinosus* (Brandt) which breeds continuously in tropical and temperate habitats (Juchault, Mocquard & Kouigan 1985; Dangerfield & Telford 1990). Although breeding is frequently discrete in woodlice, there is considerable variation in breeding phenology between species (Sutton et al. 1984; Warburg 1987), between populations (Souty-Grosset et al. 1988) and within populations between years (Dangerfield & Hassall, 1992). In many cases, the length of the breeding season is such that a female can produce up to three broods in one season (e.g. *Porcellio scaber* Latr., Davis 1978). Consequently, variation in the timing of reproduction for individual females can be inferred from these population measures (Lawlor 1976a; Sunderland, Hassall & Sutton 1976).

2.2 *Lifetime brood production*

Woodlice are potentially iteroparous between years and can produce multiple broods per season in favourable habitats (e.g *A. vulgare*, Lawlor 1976a; Miller & Cameron 1983; *P. scaber* Davis 1984; *Ligia oceanica* (Latr.), Willows 1984). Across species the number of broods per lifetime varies from one in certain populations *of A. vulgare* and other Armadillidiidae to more than six in some populations of *P. pruinosus* and *Porcellio laevis* Latr.(Warburg 1987). In laboratory cultures females frequently produce multiple broods (Merriam 1971, Dangerfield unpublished data). However, a combination of delayed maturation, short discrete breeding seasons (Willows 1984) and/or significant costs of reproduction for females (Brody, Edgar & Lawlor, 1983) can restrict normally iteroparous females to semelparity (Dangerfield & Hassall 1992).

In *P. scaber*, the number of broods produced within a season is influenced by the specific breeding phenology of the female. Females that allocate resources to reproductive events late in the season risk a reabsorption of oöytes (Davis 1978) and the loss of reproductive output if low temperatures prolong brood development (e.g. McQueen & Steel 1980). In a Californian population of iteroparous *A. vulgare*, first year females produced one brood within a season, while second year females produced two broods (Lawlor 1976a). Conversely, in a population of *A. vulgare* in East

Anglia the majority of females were semelparous, producing one brood in their second year, yet when growth conditions were favourable 19.3% of all gravid females reproduced after one year and 7.4% of females survived to reproduce after three years (Dangerfield & Hassall 1992). The potential for both iteroparous and semelparous females within a population suggests that lifetime brood production is not necessarily a fixed genetic trait as originally proposed by Lawlor (1976a), but that a range of actions are open to females in the timing and number of reproductive events.

2.3 *Fecundity and growth*

Fecundity in woodlice is positively associated with body size (several examples in Sutton et al. 1984). Large females produce larger broods than small females, hence, natality in populations is not only related to the number of reproductive females but also their size at reproduction (Rushton & Hassall 1987; Hassall & Dangerfield 1990). As woodlice show continuous growth, individual fecundity is a relatively simple trade-off between the timing of reproduction and growth rate once a minimum body size or maturation threshold (Merriam 1971) is exceeded.

Comparisons of population size structure show that even within an age class there is a considerable range in female size (e.g. Sunderland et al. 1976; Dangerfield & Telford, submitted a) due to variability in growth rate between individuals (Hubbell 1971). This variance in size within an age class (or cohort) appears to increase with time. Such a range in female size results in an equivalent range in fecundity within discrete breeding events. For example, in *P. pruinosus* within one sample of breeding females fecundity ranged from 3 to 35 (Dangerfield & Telford 1990).

The range of observed fecundity is best explained by a combination of factors that influence the growth of individuals, namely the genetic determinants of the growth exponent (Hubbell 1971), the ability of individuals to accrue resources, environmental conditions, birth date in seasonal environments, the timing of allocation of resources to reproduction and, in iteroparous individuals, timing within a temporal sequence of reproductive events. These intrinsic and extrinsic factors appear to combine in a complex way. Even in controlled laboratory conditions sibling *P. pruinosus* reared separately showed wide variation in birth mass, growth rates and size at maturity (Dangerfield & Telford, submitted b) with no predictable covariation between these measures.

2.4 *Offspring size, growth, and survivorship*

Resources devoted to a given reproductive event by a female are considered finite and therefore divisible between many small or few large offspring (Lloyd 1987). Larger offspring, assumed to have a greater proportion of parental investment, should be more likely to grow faster and have greater survivorship probabilities than smaller siblings (Levins, 1968; Smith & Fretwell, 1974; Winkler & Wallin, 1987).

In woodlice, offspring size varies significantly between species (Sutton et al. 1984), between populations as a response to moisture stress (Dangerfield & Hassall, in preparation) or female condition (Brody & Lawlor 1984) and can vary within a single population between reproductive events (Brody & Lawlor 1984; Dangerfield

& Telford 1991). This significant variation in offspring size at independence within populations is paralleled by differences in ovum size (Willows 1987; Warburg & Cohen 1992) and offspring size (Dangerfield & Telford 1990) within a brood of up to 46% of the total variation in the population (Dangerfield & Hassall, in preparation).

Female reproductive success is not only determined by the number and size of young produced, but also the proportion of these offspring that survive to reproduce themselves. Juvenile mortality, juvenile growth rates and the time to first reproduction for offspring have been measured in the context of population studies as means for given cohorts. Juvenile mortality is high, frequently up to 70% of natality after four months (Sunderland et al. 1976; Dangerfield 1987). Recent laboratory studies on *P. pruinosus* have shown that much of this mortality occurs within 10 days of independence from the marsupium and that there is only a small effect of initial offspring size on survivorship probabilities (Dangerfield & Telford, submitted b).

Variable growth rates have already been considered in the context of effects on fecundity but are also important as determinants of when young mature and, in combination with discrete breeding events, lifetime brood production. In a population of *A. vulgare* relaxation of intraspecific competition and favourable growth conditions allowed young born in July 1983 to grow rapidly and reproduce the following summer thus breaking the normal sequence of two year maturation and semelparity (Dangerfield & Hassall 1992).

3 REPRODUCTIVE FLEXIBILITY

Flexibility in female reproductive tactics are apparent in many woodlouse populations. Lifetime brood production, phenology of brood production, fecundity and offspring size all differ to some degree between individuals. The theoretical concepts of bet-hedging (Slatkin 1974) and phenotypic plasticity (Bradshaw 1965; Caswell 1989) consider such variation in reproductive traits as a strategy capable of maximising female reproductive success under changeable environmental conditions (Houston & McNamara 1992). However, in order to maximise fitness, females must first assess current environmental conditions and then respond through adaptive manipulations of resource allocation to reproduction. The implication here is that females are capable of assessing environmental conditions and can adjust resource allocation to overall reproductive patterns and individual offspring. This 'decision making' is a central tenet of current life history theory with the implicit assumption that proximate mechanisms exist to mediate such decisions.

In woodlice, some of the proximate mechanisms for such adjustments are apparent. Physiological controls on reproductive patterns occur in *A. vulgare* through induction of sexual rest in response to changes in day length (Souty-Grossett 1988), where day length is perceived as changes in light intensity (Mocquard et al. 1989). When this is combined with a maturation threshold (Merriam 1971) an effective 'decision' as to breed now or breed later occurs.

In some populations of *A. vulgare*, offspring size is negatively related to fecundity (Lawlor 1976b), and females under food stress produced large offspring (Brody & Lawlor 1984). In this case, food stress reduces growth and lowers fecundity, the consequence of which is larger offspring size. Larger offspring size would be advantageous for offspring survival in situations of reduced food availability. Although potentially adaptive (Brody & Lawlor 1984) such changes in offspring size are more an effect of a combination of reproductive traits (growth, fecundity and offspring size) than a direct 'decision' by a female. Nevertheless, this combination of factors has the same ultimate effect as a proximate mechanism that allocates more resource to each offspring during food stress.

The differential provision of nutrient to young at vitellogenesis is also a possible means of varying offspring size at independence. Willows (1987) and Warburg & Cohen (1992) have measured variance in oöcyte diameter within a brood. Any size differences in ova prior to fertilization may be enhanced during the period of development in the marsupium. Developing ova and mancas receive nutrients from brood pouch cotyledons (Hoese 1984; Hoese & Janssen 1989) and the possibility exists that brood pouch position determines the amount of nutrient uptake by individual offspring. The process of nutrient supply could be attenuated by the rate or quality of food ingested by females during brood development.

The presence of feminizing epigenetic factors in *A. vulgare* (Rigaud et al. 1991) suggest another mechanism for control of investment. In this case the females may be able to control the sex of offspring in a brood through behaviours, such as burrowing or shelter site selection, that influence temperature during brood development (Juchault & Mocquard 1988; Rigaud, Juchault & Mocquard 1991).

4 LIMITS TO FLEXIBILITY

Although flexibility and 'decision making' with respect to reproductive allocation are essential components of individual fitness in changeable environments (Via & Lande 1985) there are limits to the extent of reproductive flexibility which may prevent complete optimization of reproductive output under the current environmental conditions. The constraints of physiology, development and morphology (the organisms 'state' sensu McNamara & Houston 1986; Mangel & Clark 1988) delimit the range of actions possible in response to the environment. In woodlice, many of these constraints are a function of body size and body size x environment interactions.

We have already noted that fecundity is a positive function of female size. The volume of the marsupium of a given female is related to her size and is finite. Given the assumption that there is a minimum offspring size at independence necessary for offspring survival (Dangerfield & Telford, submitted b), maximum fecundity becomes limited by brood pouch volume. Decisions on allocation of resources to fecundity are thus constrained by an upper limit.

Another potential morphological constraint is the number of cotyledons in the marsupium which is known to vary between species (Lewis 1990). An increase in cotyledon number will result in greater surface area for nutrient transfer, as nutrients are supplied to from the cotyledons (Hoesse & Janssen 1989) developing mancas may be limited by brood pouch morphology. Similarly, overall brood development rates may also be affected by rate of nutrient supply. The presence of cotyledons also reduce the volume of the marsupium, which suggests a potential trade-off between nutrient supply and maximal fecundity.

Flexibility in reproductive allocation is also limited by constraints on resource acquisition. Woodlice are dependent on adequate moisture supply and have developed complex behavioural and physiological mechanisms to maintain water balance (Edney 1954; Cloudsley-Thompson 1975; Warburg 1987). These mechanisms frequently restrict foraging which can reduce food intake and resources available for reproduction.

Flexibility will also be limited by the specific combination of genes inherited from recent and distant evolutionary past (Harvey & Pagel 1991). A single evolutionary origin for this group (Schmalfuss 1989) would suggest a conservative genome. These phylogenetic considerations moderate the extent of possible variation in reproductive traits as many physiological and morphological constraints will be determined by the genotype.

5 FLEXIBILITY AND INDIVIDUAL FITNESS

Heterogeneity in habitat conditions and resource availability, biotic interactions and genetic variation dictate that a population will be made up of a number of different phenotypes. Each individual will be unique with respect to the conditions it has experienced and the resource it has obtained. Through the lifetime of an individual the detrimental effects of conditions must be neutralised by the allocation of resources to maintenance, only then can any excess resource be partitioned between growth and reproduction. The timing and trade-off in resource allocation between growth and reproduction is potentially unique to the individual. Recently, this distinction between an average, population level, suite of reproductive strategy and specific tactics employed by individuals has been made more explicit in a series of life history models that are focused on individual rather than group fitness (Charnov & Berrigan 1991; Schultz 1991; Yoshimura & Clark 1991; Houston & McNamara 1992).

Coarse scale habitat effects, when combined with the morphological and physiological capacities of the organism, will produce a broad range of life history responses at the population level. In woodlice, the effects of temperature (McQueen 1976; Hubbell 1971), shelter site availability (Paris 1963) and food quality (Rushton & Hassall 1987) on growth rate may be central to this broad scale effect (Dangerfield & Hassall 1992). However, what is optimal for the average individual in the population (arithmetic means of various life history traits) may not maximise fitness for individuals with greater or less than average resource acquisition. As the environment or the effects of the environment on resource acquisition become more variable in time or space, a larger proportion of individuals will fall into these latter categories within the lifetime of the organism and an ability to respond to stress or abundant resource through flexible allocation to reproduction would be selected. Yoshimura & Clark (1991) have taken this further with the idea that 'worst case' conditions may produce greater selection pressures to avoid total reproductive failure than to increase reproductive output when conditions are favourable.

The wide range of individual tactics seen between woodlouse populations and the evidence from studies within populations suggest that such flexibility in individual decisions on reproduction exists. However, focused observations and critical tests are necessary to establish the extent of reproductive flexibility in different populations and to address the assumption that such flexibility does indeed enhance fitness.

6 OBSERVATIONS AND TESTABLE HYPOTHESES

The advantage of terrestrial isopods as study organisms is evident in the wide range and depth of basic biological knowledge available for this group (see overview by Hassall 1991). Here we have summarised the extensive variation in reproductive traits that has been observed in field and laboratory populations and considered these as an adaptive response by individuals to current environmental conditions (but see McGinley, Temme & Geber (1987) for a non-adaptive explanation for phenotypic variance). This concept of adaptive phenotypic plasticity (Caswell 1989) as a suite of apparent resource allocation 'decisions' by females could be experimentally tested with woodlice as they represent a readily manipulatable system in which individual fitness can be quantified.

We conclude with examples of questions (Q) answered by direct observations and testable hypotheses (H) for each of the main life history traits associated with reproduction that we see as important for quantitative tests of reproductive flexibility as a strategy to maximize fitness. For convenience we follow the sequence of subheadings in Section 2 and provide a brief description of the type of experiment and/or observations necessary. Further details of the suggested experimental protocols are available on request from the authors.

6.1 *Breeding phenology*

Q1 *Does the breeding phenology within a population vary between individuals?*
The suggestion that combinations of reproductive patterns exists within populations (Dangerfield & Hassall 1992) requires more explicit delimitation of reproductive events for individuals. As Stearns (1976) noted, there is a need to establish for individuals what has to date been recorded only within populations. Observations of known individuals is difficult within natural populations but could be achieved in the laboratory with slight modifications to established culture and marking techniques.

H1 *Individual breeding phenology is genetically determined.*
Direct genetic control (De Jong 1990) or a genetically determined norm of reaction (Stearns & Koella 1986) may exist for reproductive traits such as the timing of reproduction. If this is the case under constant conditions, females with similar genotypes, and most importantly their offspring, should reproduce at the same time window after independence. Heritability experiments can be designed to test this hypothesis which specifically addresses the existence of phenotypic plasticity. Such experiments could be completed in two years under laboratory conditions for a range of species, although the use of fast growing porcellionids such as *P. scaber* or *P. pruinosus* would be recommended.

H2 *The change from semelparity to iteroparity is determined by specific combinations of environmental conditions.*

Manipulations of laboratory environments could be carried out to determine the transition point between semelparous and iteroparous reproduction for females. A factorial experiment in which temperature, moisture and food availability could be carried out on a sub-divided laboratory population. Even though replication would need to be extensive this experiment is considered feasible over a three year period. Alternatively a reciprocal transplant experiment could be carried out on field populations.

Q2 *What temporal patterns of iteroparity exist within populations?*
Results from experiments carried out to test H2 can be used to predict intraspecific patterns of iteroparity in selected field populations.

Q3 *Is brood development time related to the number of marsupium cotyledons?*
Comparative observations of brood development time under controlled conditions (temperature, food availability, female age, first brood) would assess the role of cotyledon number in brood maturation. Covariance analyses would control for effects of fecundity. Post-brood release dissection of females would establish if cotyledon number or cotyledon size explains a significant proportion of intraspecific variation in development time.

6.2 *Growth and fecundity*

H3 *Growth rate is under genetic control.*
Experiments spanning at least two generations, or two to three years for most temperate species, in which the growth rates of individuals, reared in isolation under constant environmental conditions, are measured (see Dangerfield & Telford, submitted b) followed by controlled mating between individuals of known relative growth rates would assess the heritability of 'fast' or 'slow' growth.

H4 *Resource availability has no effect on fecundity in similar sized females.*
Rearing experiments of known individuals under different conditions of resource availability and measurement of fecundity at first reproduction and subsequent reproductive events would determine the effects of resource allocation to brood size. The assumption being that observed residual variation in fecundity after taking into account the effects of female size is due to resource allocation.

6.3 *Offspring size*

H5 *Nutrient supply to mancas in the marsupium does not limit offspring dry mass at independence.*
Experimental ablation of ova on release from ovary into the marsupium followed by measurements of offspring dry mass at independence would assess the idea that nutrient supply to mancas limits offspring size at independence. Such a study is possible with the larger porcellionid and armadillid species and would benefit from a multi-species comparison.

Q4 *Does the position of individual mancas in the marsupium influence size at independence?*
Careful disSection of females close to brood release together with live and dry mass estimates of manca size will determine if position, particularly with respect to cotyledons, explains the observed variation in offspring size at independence. These observations consider the assumption that mancas closest to cotyledons receive a disproportionate share of available nutrients during development, particularly as the greater proportion of mass gain in mancas occurs during the latter quarter of brood development (Lawlor 1976b; Ma et al. 1990).

Q5 *Is there differential investment in newly fertilized ova?*
Observations by Warburg & Cohen (1992) and Willows (1984) suggest that there is considerable variation in ovum development within the ovary. The comparison between variances in allocation to ova and size at independence would address the question of when females begin to vary investment in individual offspring. A significant proportion of variation in investment at vitellogenesis would be more consistent with theories of bet-hedging (Slatkin 1974), whilst differential development within the brood pouch would be more consistent with concepts of phenotypic plasticity (Caswell 1989).

6.4 *Offspring growth and survivorship*

H6 *Environmental conditions determine the minimum offspring size within broods.*
The concept of a minimum size threshold below which offspring survival probabilities are much reduced (Dangerfield & Telford, submitted b) is central to the trade-off between female size and fecundity and may also explain differences in offspring size between populations. Rearing of females from different populations and measurement of offspring dry mass at independence under various conditions would assess the flexibility of female allocation to offspring. Although experiments on grwoth and survivorship of individually reared offspring demand a significant time commitment they are feasible and would be valuable if carried out on different species, preferea-bly using females from sympatric populations.

H7 *Offspring size is independent of resource quality supplied to females during brood development.*
A more detailed manipulation of the type described by Brody & Lawlor (1984) in which females from several species of similar size, age and reproductive condition are fed a range of foods and offspring size determined would test this hypothesis.

6.5 *Flexibility*

H8 *Environmental perturbations have no influence on lifetime offspring production of females.*
Rearing of females under laboratory conditions with discrete perturbations of food availability, temperature, shelter site availability and measurements of breeding phenology and reproductive output would answer the question of the ability of females to be flexible toward rapid changes in environmental conditions. If females do re-

spond with changes in resource allocation to reproduction then implicit is a perception of the prevailing conditions prior to any given breeding event. This is also an indirect assessment of non-genetic constraints (sensu Atkinson 1984) to reproduction.

Q6 *Do populations that contain individuals with high levels phenotypic plasticity also have low genetic variance?*
Observations of phenotypic plasticity would be combined with measures of genetic variance (for example gel electrophoresis of soluble enzymes). The assumption here is that, over time, conditions that produce strong selection pressure for phenotypic plasticity would result in reduced overall genetic variance if phenotypic plasticity is a heritable trait. In effect, a population would be invaded by the genotype that allows for flexibility. A comparison of populations at different locations within biogeographic range of the species would assess the additional idea that populations at the edge of the species range should contain higher proportions of individuals with flexibility.

7 CONCLUSIONS

The impact of environmental variance and stochastic events on the evolution of life histories through individual variation has only recently been considered theoretically (e.g. Schultz 1991; Yoshimura & Clark 1991). The conceptual elegance of these ideas rests on the critical assumption that proximate mechanisms exist which allow 'decision making' by individuals in their allocation of resources to reproductive events. Several such mechanisms that give an individual reproductive flexibility in response to current conditions have already been identified in woodlice. The existence of these mechanisms, the wide range of observed life history tactics among and within species and the versatility of woodlice as experimental animals make woodlice an ideal model for quantitative tests of evolutionary theory.

ACKNOWLEDGEMENTS

We thank Bruce Riddoch for helpful discussions and comments on an earlier version of this paper and an anonymous reviewer for helpful suggestions.

REFERENCES

Atkinson, D. 1985. Information, non-genetic constraints and the testing of theories of life history variation. In: R.M. Sibly & R.H. Smith (eds.), *Behavioural ecology: ecological consequences of adaptive behaviour.* Symposia of the British Ecological Society 25. Oxford: Blackwell.

Bradshaw, A.D. 1965. Evolutionary significance of phenotypic plasticity in plants. *Adv. Genet.* 13: 115-55.

Brody, M.S. & L.R. Lawlor. 1984. Adaptive variation in offspring size in the terrestrial isopod *Armadillidium vulgare. Oecologia* 61: 55-59.

Brody, M.S. M.H. Edgar & L.R. Lawlor. 1983. A cost of reproduction in a terrestrial isopod. *Evolution* 37: 653-655.

Caswell, H. 1983. Phenotypic plasticity in life-history traits: demographic effects and evolutionary consequences. *Amer. Zool.* 23: 35-46.

Caswell, H. 1989. Life-history strategies. In. J.M. Cherret (ed.), *Ecological Concepts: The contribution of ecology to an understanding of the natural world.* Symposia of the British Ecological Society 29. Oxford: Blackwell.

Charnov, E.L. & D. Berrigan. 1991. Evolution of life history parameters in animals with indeterminate growth, particularly fish. *Evol. Ecol.* 5: 63-68.

Cloudlsey-Thompson, J.L. 1975. Adaptations of Arthropoda to arid environments. *Ann. Rev. Entomol.* 20: 261-283.

Dangerfield, J.M. 1987. *Variation in the life history and population parameters of Armadillidium vulgare (Latreille).* Unpublished Ph D. thesis: University of East Anglia, Norwich

Dangerfield, J.M. & M. Hassall. 1992. Phenotypic variation in the breeding phenology of the woodlouse *Armadillidium vulgare. Oecologia* 89: 140-146.

Dangerfield, J.M. & S.R. Telford. 1990. Breeding phenology, variation in reproductive effort and offspring size in a tropical population of the woodlouse *Porcellionides pruinosus. Oecologia* 82: 251-258.

Dangerfield, J.M. & S.R. Telford submitted a. Population size structure and sex ratios from some woodlice (Crustacea: Oniscidae) in southern Africa. *J. Trop. Ecol.*

Dangerfield, J.M. & S.R. Telford submitted b. Offspring growth and survivorship in the woodlouse *Porcellionides pruinosus. Ethology, Ecology & Evolution.*

Davis, R.C. 1978. *Ecological studies of Isopoda and Diplopoda in dune grassland.* Unpublished PhD Thesis: University of Leeds

Davis, R.C. 1984. Effects of weather and habitat structure on the population dynamics of isopods in a dune grassland. *Oikos* 42: 387-395.

De Jong, G. 1991. Genotype-by-environment interaction and the genetic covariance between environments: multilocus genetics. *Genetica* 81: 171-177.

Edney, E.B. 1954. Woodlice and the land habitat. *Biol. Rev.* 29: 185-219.

Harvey, P.H. & M.D. Pagel 1991. *The comparative method in evolutionary biology.* Oxford: University Press.

Hassall, M. 1991. Interdisciplinary approaches to research on terrestrial isopods. In: P. Juchault & J-P. Mocquard (eds.), *The Biology of Terrestrial Isopods III.* Proceedings of the Third International Symposium on the Biology of Terrestrial Isopods. Poitiers: Universite de Poitiers.

Hassall, M. & J.M. Dangerfield 1990. Density-dependent processes in the population dynamics of *Armadillidium vulgare* (Isopoda: Oniscidae). *J. Anim. Ecol.* 59: 941-958.

Hassall, M. & J.M. Dangerfield submitted. Effects of habitat and climate on the population dynamics of Armadillidium vulgare (Isopoda: Oniscidae). *J. Anim. Ecol.*

Hoese, B. 1984. The marsupium in terrestrial isopods. In: S.L. Sutton & D. Holdich (eds.), *The Biology of Terrestrial Isopods.* Symposium of the Zoolological Society of London 53. Oxford: Clarendon.

Hoese, B. & H.H. Janssen 1989. Morphological and physiological studies on the marsupium in terrestrial isopods. *Monit. Zool. Ital. N.S. Monogr.* 4: 153-157.

Houston, A.I. & J.M. McNamara 1992). Phenotypic plasticity as a state-dependent life-history decision. *Evol. Ecol.* 6: 243-253.

Hubbell, S.P. 1971. Of sowbugs and Systems: the ecological bioenergetics of a terrestrial isopod. In: B.C. Patten (ed.), *Systems Analysis and Simulation in Ecology.* New York: Academic Press.

Juchault, P. & J-P. Mocquard 1988. Effects de la temperature sur le sex ratio de la descendance et la physiologie sexuelle des femelles *d'Armadillidium vulgare* Latr. (Crustacea, Oniscidea) hebergeant une bacterie feminisante. *C. R. Acad. Sci. Paris,* 306(III): 321-324

Juchault, P., J-P. Mocquard & S. Kouigan 1985. Etude experimentale de l'influence des facteurs externes (temperature et photoperiode) sue le cycle de reproduction du crustace oniscoide *Porcellionides pruinosus* (Brandt) provenant de populations Africaine (Togo) et Europeenne (France). *Crustaceana* 48: 307-315.

Lawlor, L.R. 1976a. Moulting, growth and reproductive strategies in the terrestrial isopod *Armadillidium vulgare. Ecology* 57: 1179-1194.

Lawlor, L.R. 1976b. Parental investment and offspring fitness in the terrestrial isopod *Armadillidium vulgare. Evolution* 30: 775-785.

Lewis, F. 1990. The relationship between broodpouch cotyledons, aridity and advancement. In: P. Juchault & J-P. Mocquard (eds.), *The Biology of Terrestrial Isopods III.* Proceedings of the Third International Symposium on the Biology of Terrestrial Isopods. Poitiers: Universite de Poitiers.

Lloyd, D.G. 1987. Selection of offspring size at independence and other size-versus-number strategies. *Amer. Nat.* 129: 800-817.

Ma, H.H.T., P.K.S. Lam & D. Dudgeon 1991. Inter- and intraspecific variation in the life histories of three sympatric isopods in a Hong Kong forest. *J. Zool. (Lond)* 224: 677-687.

Mangel, M. & C.W. Clark 1988. *Dynamic modelling in behavioural ecology.* Princeton: Princeton University Press.

McGinley, M.A., D.H. Temme & M.A. Geber 1987. Parental investment in variable environments: Theoretical and empirical considerations. *Amer. Nat.* 130: 370-398.

McQueen, D.J. 1976. The influence of climatic factors on the demography of the terrestrial isopod *Traechoniscus rathkei* Brandt. *Can. J. Zool.* 54: 2185-2199.

McQueen, D.J. & C.G.H. Steel 1980. The role of photoperiod and temperature in the initiation of reproduction in the terrestrial isopod *Oniscus asellus* Linnaeus. *Can. J. Zool.* 58: 235-240.

McNamara, J.M. & A.I. Houston 1986. The common currency for behavioural decisions. *Amer. Nat.* 127: 358-378.

Merriam, H.G. 1971. Sensitivity of terrestrial isopod populations (*Armadillidium*) to food quality differences. *Can. J. Zool.* 49: 667-674.

Mocquard, J-P., P.J. Juchault & C. Souty-Grosset 1989. The role of environmental factors (temperature and photoperiod) in the reproduction of the terrestrial isopod *Armadillidium vulgare* (Latreille, 1804). *Monit. Zool. Ital. N.S. Monogr.* 4: 455-475.

Miller, R.H. & G.N. Cameron 1983. Intraspecific variation of life history parameters in the terrestrial isopod *Armadillidium vulgare. Oecologia* 57: 216-226.

Paris, O.H. & F.A. Pitelka 1962. Population characteristics of the terrestrial isopod *Armadillidium vulgare* in California grassland. *Ecology* 43: 229-248.

Rigaud, T., P. Juchault & J-P. Mocquard 1991. Experimental study of temperature effects on the sex ratio of broods in terrestrial Crustacea *Armadillidium vulgare* Latr. Possible implications in natural populations. *J. Evol. Biol.* 4: 603-617.

Rigaud, T., C. Souty-Grosset, R. Raimond, J-P. Mocquard & P. Juchault 1991. Feminizing endocytobiosis in the terrestrial Crustacean *Armadillidium vulgare* Latr. (Isopoda): Recent acquisitions. *Endocytobiosis & Cell Res.* 7: 259-273.

Rushton, S.P. & M. Hassall 1987. Effects of food quality on isopod population dynamics. *Func. Ecol.* 1: 359-368

Schmalfuss, H. 1989. Phylogenetics in Oniscidae. *Monit. Zool. Ital. N.S. Monogr.* 4: 3-27.

Schultz, D.L. 1991. Parental investment in temporally varying environments. *Evol. Ecol.* 5: 415-427.

Sibly, R.M. & P. Calow 1986. *Physiological ecology of animals.* An evolutionary approach. Oxford: Blackwell Scientific Publications.

Slatkin, M. 1974. Hedging one's evolutionary bets. *Nature* 250: 704-705.

Smith, C.C. & D.S. Fretwell 1974. The optimal balance between size and number of offspring. *Amer. Nat.* 108: 499-506.

Souty-Grosset, C., A. Chentoufi, J-P. Mocquard & P. Juchault 1988. Seasonal reproduction in the terrestrial isopod *Armadillidium vulgare* (Latreille): Geographic variability and genetic control of the response to photoperiod and temperature. *Invert. Rep. Dev.* 14: 131-151.

Southwood, T.R.E. 1977. Habitat, the template for ecological strategies? *J. Anim. Ecol.* 46: 337-365.

Stearns, S.C. 1976. Life history tactics: a review of the ideas. *Q. Rev. Biol.* 51: 3-47.

Stearns, S.C & J.C. Koella 1986. The evolution of phenotypic plasticity in life-history traits: predictions of reaction norms for age and size at maturity. *Evolution* 40: 893-913.

Sunderland, K.D., M. Hassall & S.L. Sutton 1976. The population dynamics of *Philoscia muscorum* (Crustacea, Oniscoidea) in a dune grassland ecosystem. *J. Anim. Ecol.* 45: 487-506.

Sutton, S.L., M. Hassall, R. Willows, R.C. Davis, A. Grundy & K.D. Sunderland 1984. Life histories of terrestrial isopods. In: S.L. Sutton & D. Holdich (eds.), *The Biology of Terrestrial Isopods*. Symposium of the Zoolological Society of London 53. Oxford: Clarendon.

Via, S. & R. Lande 1985. Genotype-environment interaction and the evolution of phenotypic plasticity. *Evolution* 39: 505-522.

Warburg, M.R. 1987. Isopods and their terrestrial environment. *Adv. Ecol. Res.* 17: 187-242.

Warburg, M.R. & N. Cohen 1992. Reproductive pattern, allocation and potential of an iteroparous isopod from a xeric habitat in the Mediterranean region. *J. Arid Environ.* 22: 161-171.

Warburg, M.R., K.E. Linsenmair & K. Bercovitz 1984. The effect of climate on the distribution and abundance of isopods. In: S.L. Sutton & D. Holdich (eds.), *The Biology of Terrestrial Isopods*. Symposium of the Zoolological Society of London 53. Oxford: Clarendon.

Willows, R.I. 1984. Breeding phenology of woodlice and oostegite development in *Ligia oceanica* (L.) (Crustacea). In: S.L. Sutton & D. Holdich (eds.), *The Biology of Terrestrial Isopods*. Symposium of the Zoolological Society of London 53. Oxford: Clarendon.

Willows, R.I. 1987. Intrapopulation variation in the reproductive characteristics of two populations of *Ligia oceanica* (Crustacea: Oniscidae). *J. Anim. Ecol.* 56: 331-340.

Winkler, D.W. & K. Wallin 1987. Offspring size and number: a life history model linking effort per offspring and total effort. *Amer. Nat.* 129:708-720

Yoshimura, J. & W.M. Shields 1987. Probabilistic optimization of phenotypic distributions: a general solution for the effects of uncertainty on natural selection? *Evol. Ecol.* 1: 125-138.

Yoshimura, J. & C.W. Clark 1991. Individual adaptations in stochastic environments. *Evol. Ecol.* 5: 13-192.

Isopod distribution at different scaling levels

E. Hornung
Department of Ecology, JATE University, Hungary

M.R. Warburg
Department of Biology, Technion, Haifa, Israel

ABSTRACT

The isopod fauna was studied in different habitats within the Mediterranean region of Northern Israel. Four habitats were involved in the survey, two grasslands, an oak-woodland and a planted pine forest. All isopod specimens were methodically hand-collected from under stones within a randomly chosen area of 400 m^2 divided into sixteen 5 × 5 m sized quadrats. The sampling took place during winter and spring when the isopods were at their peak activity period. The number of stones in each quadrat and their approximate size were also recorded. The data were evaluated at three scaling levels: 1) Macrodistribution – comparing the species composition and abundance of isopods among different habitats within the studied geographical region. 2) Microdistribution – comparison of isopod distribution among different quadrats within the same habitat. 3) Minidistribution – cohabitation of isopod species under single stones within a habitat. Isopods were unevenly distributed among the habitats studied. Some species were more abundant in grassland (*Porcellio ficulneus*, *Schizidium tiberianum*, *Armadillo* sp. 'brown'), whereas others (*Armadillo officinalis*) were found in large numbers in planted pine forest. Only two species (*Porcellionides pruinosus* and *Chaetophiloscia* spp. complex) were abundant in all habitats. There was a negative relationship between isopod aggregation and the number of stones within a habitat. Medium and large stones proved to be the most suitable shelters for isopods.

1 INTRODUCTION

The distribution of isopod species has been investigated mainly on a zoogeographical scale. The large scale distribution pattern of Mediterranean isopod species in the genera *Chaetophiloscia*, *Schizidium* and *Porcellio* was recently elaborated by Schmalfuss (1988, 1991, 1992). Previous studies on the distribution patterns of the isopod species in this region (Warburg et al. 1978) showed that they all are widely distributed in the Mediterranean region of northern Israel (Fig. 1b-d). Such large scale regional zoogeographical surveys suggest that several isopod species occur uniformly throughout the region.

Only a few studies have dealt with the distribution of isopods on a smaller scale, that is, their presence in different habitats within the same region. Thus Herold (1930) studied the distribution of terrestrial isopods in the Baltic region, as related to the ecological conditions prevailing within their habitats. Miller (1938) showed a zonation and succession pattern of distribution in thirteen Californian isopod species. Similarly, Beyer (1964) described the distribution of eighteen isopod species in different habitats in Central Germany. These studies were largely qualitative, with little supporting quantitative data.

The isopod species composition in several habitats within the Mediterranean region of Israel has been previously described (Warburg et al. 1978). In that study the isopod fauna was compared in eight habitats ranging from semi-arid grassland to mesic oak-woodland. A few species were common to the moist habitats. Similarly, Beyer (1964) showed that whereas some isopod species had restricted distributions, others had wider distributions. In the Somali coast, four isopod species showed a pronounced zonation near the sea shore; five other species were widely distributed in terrestrial habitats (Chelazzi & Ferrara 1978).

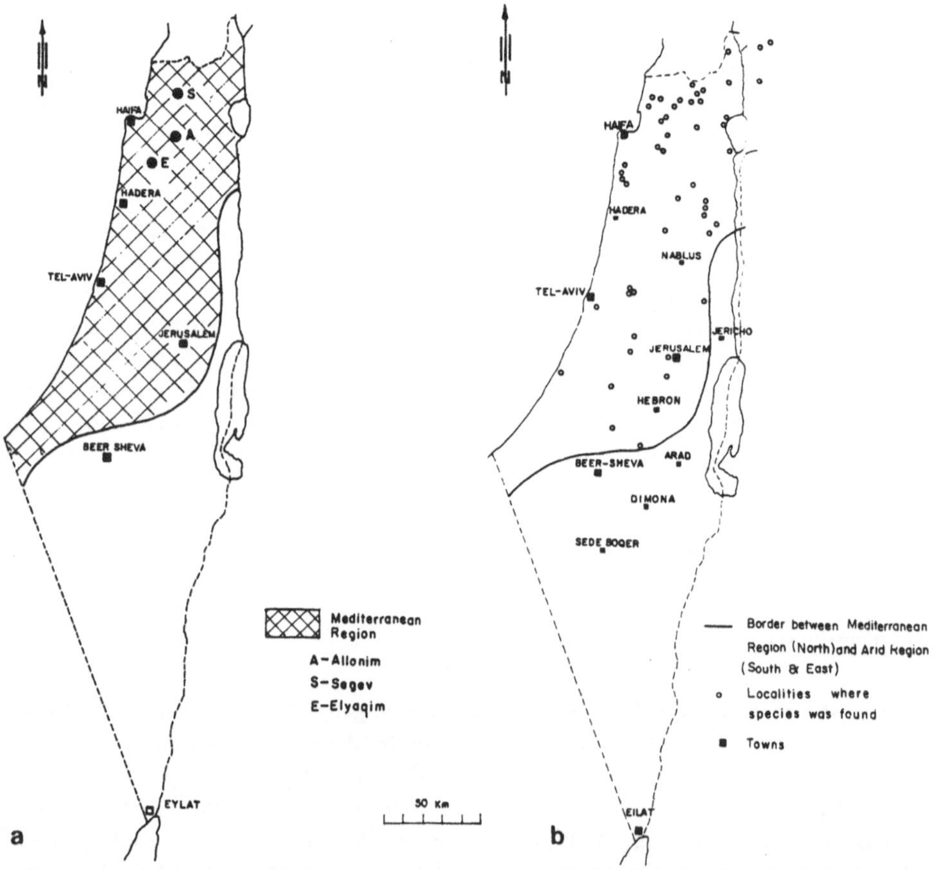

Figure 1. a) Map of Israel showing the Mediterranean region (hatched area) and the localities of sampling areas (heavy dots); b-d) Known distribution of frequent species in Israel (b = *Porcellio ficulneus*).

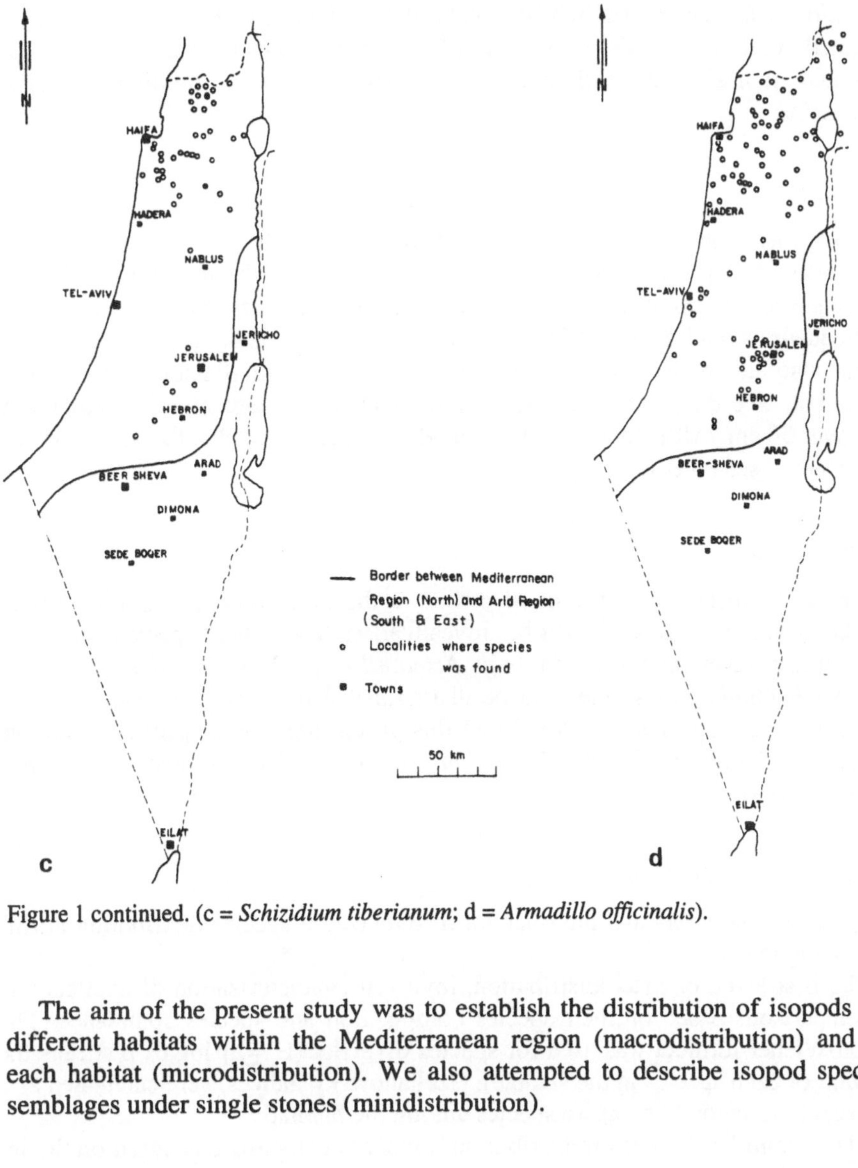

Figure 1 continued. (c = *Schizidium tiberianum*; d = *Armadillo officinalis*).

The aim of the present study was to establish the distribution of isopods in four different habitats within the Mediterranean region (macrodistribution) and within each habitat (microdistribution). We also attempted to describe isopod species assemblages under single stones (minidistribution).

2 MATERIALS AND METHODS

2.1 *Localities and habitats*

Four habitats were selected at three localities. The localities were located in the vicinity of the villages Elyaqim, Segev and Allonim (Fig.1a). One of the habitats was a grassland at Elyaqim, located in the southern part of Mt.Carmel at 300 m elevation, with an average annual rainfall of about 600 mm. The area of this grassland was about 100 km^2. A second grassland area was chosen at Segev in the Lower Galil Mts., at 550 m elevation and with about 650 mm rainfall. This grassland was a large patch (about 150 × 500 m) in the third habitat, a 35 year-old planted pine forest

(*Pinus halopensis*). The fourth habitat, a natural oak-woodland (*Quercus calliprinos*) was located at Allonim, in the Lower Galil Mts., at an elevation of 200 m, with 650 mm annual rainfall. The soil type of all habitats was a mixture of terra-rossa/rendzina.

2.2 *Sampling*

Field work was done in winter and spring (February, March) which is the main activity season of isopods in this region. Specimens were collected from under the stones in an area of 400 m^2 at each locality. The areas were each divided into sixteen, 5 × 5 m quadrats randomly distributed within a homogeneous habitat of several km^2. We listed also the number and estimated size class (maximum diameter) of stones in each quadrat. The three size categories were: small = 5-20 cm; middle = 20-50 cm; large = over 50 cm. All stones and the isopod assemblages under the single stones were registered separatedly.

2.3 *Identification*

The large sized specimens could be identified in the field whereas the minute ones were taken to the laboratory for further investigation. There were species which are recently under taxonomic revision (e.g. *Armadillo* sp. 'brown', *Chaetophiloscia* spp.). As *Chetophiloscia* species can be distinguished only by their males, we referred to them as a species complex. From this group three valid species are known to occur in Northern Israel: *C.elongata aharoni*, *C.lagoi* and *C.warburgi* (Schmalfuss 1991).

2.4 *Scale-related distribution*

We established the inter- and intra-habitat differences in species distribution at different scaling levels.

1) The first level, or macrodistribution, involved characterization of habitats and their comparisons based on their species composition and species abundance. The Shannon-Wiener formula was used for species diversity ($H' = \Sigma p_i \log_2 p_i$; p_i means the proportion of each species in the sample). Czekanowsky indices were calculated as a beta diversity function to compare species among the habitats.

2) The second level, or microdistribution, involved comparisions based on the individual quadrats within habitats, stating the dispersion indices of isopods and the correlation of isopod abundance with the stone coverage in each sampling unit (quadrat). Simple dispersion index was calculated in order to establish dispersion type ($I_D = s^2/\bar{x}$; s^2 is the variance and $\bar{x}^2$ the mean of samples). Regression analysis and its correlation coefficients were obtained through Microstat statistical computer program.

3) The third level, or minidistribution, involved comparisons of species association based on cohabitation of different species in the same shelter, that is under the same stone. Chi^2 tests were used to evaluate species associations.

3 RESULTS

3.1 *Macrodistribution*

3.1.1 *Species composition*

During the study period we found at least 12 species of isopods that inhabit the Mediterranean region of Northern Israel (Table 1). Six of these were relatively more abundant, while the remaining six were found occasionally, appearing in very low frequency (1-5 specimens). Two of the abundant taxa (*Porcellionides pruinosus*, *Chaetophiloscia* species complex), were found in all four habitats, one species (*Schizidium tiberianum*) was collected in three habitats and two more species were found only in two habitats (*Armadillo officinalis* and *Armadillo* sp.). There were species which occured only in grasslands (*Porcellio ficulneus, Armadillo* sp.) or exclusively in woodland habitats (*A.officinalis*).

3.1.2 *Species richness and diversity*

The number of species per habitat did not differ much. In three of the four habitats studied 6 species were found whereas in one habitat 8 species were found. Diversity indices ranged between 0.35 to 0.66 (Table 1). The highest diversity was in Allonim (0.66), followed by Elyaqim (0.58). The two Segev habitats showed a low diversity (0.35, 0.49, respectively).

3.1.3 *Inter-habitat similarities*

The values of Chekanowsky indices refer to the similarities of habitats based on

Table 1. Species composition, abundance and diversity of isopods in four habitats of the Mediterranean region (Northern Israel, 1992).

Species	Habitats							
	Elyaqim grass		Segev grass		Segev forest		Allonim forest	
	No	%	No	%	No	%	No	%
**Trichoniscus* sp.	0	0	0	0	0	0	1	0.46
Bathytropa wahrmani	0	0	0	0	2	0.26	0	0
**Chaetophiloscia* spp.	67	5.61	58	11.74	110	14.38	87	39.73
Porcellio ficulneus	530	44.3	0	0	0	0	0	0
Porcellionides pruinosus	345	28.85	241	48.78	41	5.36	49	22.37
**Proporcellio* sp.	2	0.17	10	2	0	0	3	1.37
**Leptotrichus* spp.	0	0	7	1.42	10	1.31	4	1.83
Nagurus carinatus	0	0	3	0.61	0	0	0	0
Armadillo officinalis	0	0	0	0	583	76.21	34	15.53
**Armadillo* sp. 'brown'	77	6.44	175	35.42	0	0	0	0
Armadillo tuberculatus	0	0	0	0	0	0	5	2.28
Schizidium tiberianum	175	14.63	0	0	19	2.48	36	16.44
total	1196		494		765		219	
%		44.89		18.54		28.72		8.22
No. spp	6		6		6		8	
H (S)	0.58		0.49		0.35		0.66	

*Species under taxonomic revision.

species composition and abundance. The highest similarity was found between the two grassland (0.45) and the two forest habitats (0.38), respectively. The value of Chekanowsky indices were the lowest in the case of grassland-forest comparisions: 0.13 (Elyaqim-Segev pine forest) and 0.17 (Segev grassland-Segev pine forest).

3.1.4 *Isopod abundance*

During the course of this study we collected 2674 individuals. The habitat of Elyaqim grassland was the richest in isopod specimens (44.89% out of the total isopods found in the four habitats), whereas Allonim was the poorest in this respect (8.22%) (Table 1). The two Segev habitats contained nearly the similar number of isopod specimens (Fig. 2).

A.officinalis comprises the majority of isopod specimens in Segev pine forest, about 76% of the whole isopod community. In the other habitats more than one predominant species occured: *P.ficulneus* and *P.pruinosus* in Elyaqim; *P.pruinosus* and *Armadillo* sp. in Segev grassland, whereas in Allonim all the 4 abundant species had nearly similar frequencies (Fig. 3).

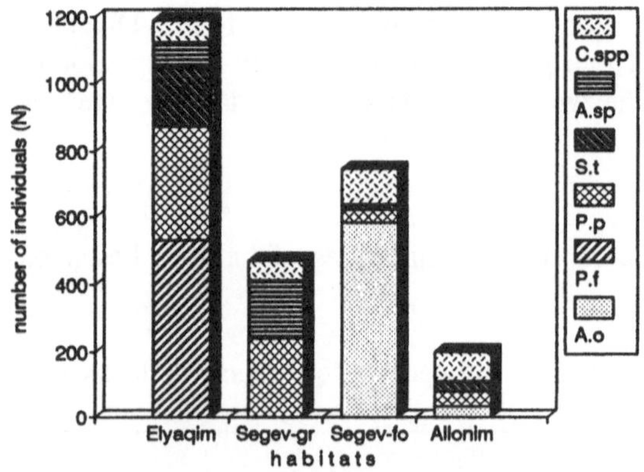

Figure 2. Composition of isopod communities and total numbers of abundant species in the four habitats (Abbreviations: C.spp. = *Chaetophiloscia* species complex; A.sp. = *Armadillo* sp. 'brown'; S.t = *Schizidium tiberianum*; P.p = *Porcellionides pruinosus*; P.f = *Porcellio ficulneus*; A.o = *Armadillo officinalis*; Segev-gr = Segev grassland; Segev-fo = Segev forest).

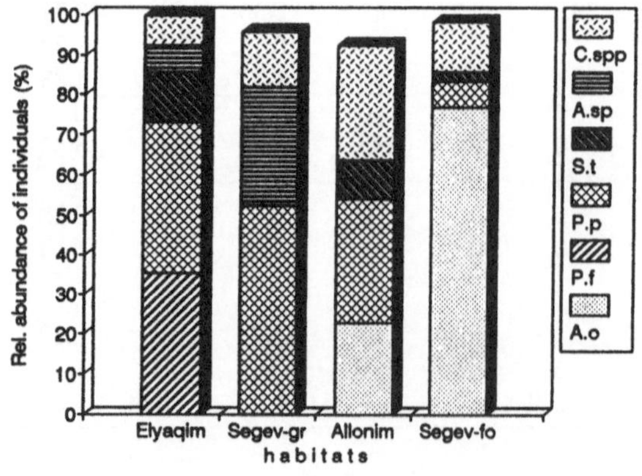

Figure 3. Relative frequency of abundant species in the four habitats (for abbreviations see Fig. 2.).

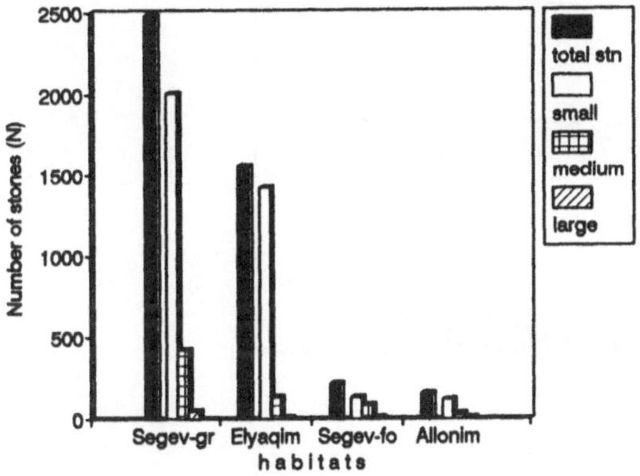

Figure 4. Total numbers and distribution of different sized stones in the studied habitats (numbers/400m^2).

Table 2. Numbers and percentages of stones inhabited by isopods.

	Total no. of specimens	Total no. of stones	Total no. of occupied stones	% of occupied stones
Segev-gr	494	4012	351	8.75
Elyaqim	1196	3396	457	13.46
Allonim	219	350	106	30.29
Segev-fo	765	291	179	61.51

3.1.5 *Stone coverage.*

As the main technique used in this study was based on collecting isopod specimens from under the stones, the number and size of the stones providing shelters are given for the four habitats. The greatest number of stones was found in the two grassland habitats (Segev-grassland and Elyaqim). Most of these stones were of small size, less than 20 cm. The two woodland habitats (Segev-forest and Allonim) had a rocky type of underground that contained fewer, but mainly bigger, stones (Fig. 4).

Isopods used the shelter of stones differently in the different habitats. The number of stones and the rate of their use was inversely related: a considerably lower percentage of stones was occupied by isopods in grasslands, where the total stone number was the highest (Table 2). Isopods were never found under stones smaller than 5 cm or on the ground surface during daytime.

3.2 *Microdistribution*

3.2.1 *Distribution of species and specimens.*

The distribution and abundance of isopods shows that different plots within a habitat differ in their species composition. Furthermore, the distribution of both species and specimens is not even within the habitat (Fig. 5). Calculation of dispersion indices based on quadrat data enables us to assess the type of species distribution within each habitat (Table 3). Some isopod species show a tendency to aggregate regardless of their habitats (*S.tiberianum, Armadillo* sp.). In other species their dispersion pattern varies among habitats (*Chaetophiloscia* spp., *P.pruinosus*). In the two grassland habitats they do not tend to aggregate (Table 3).

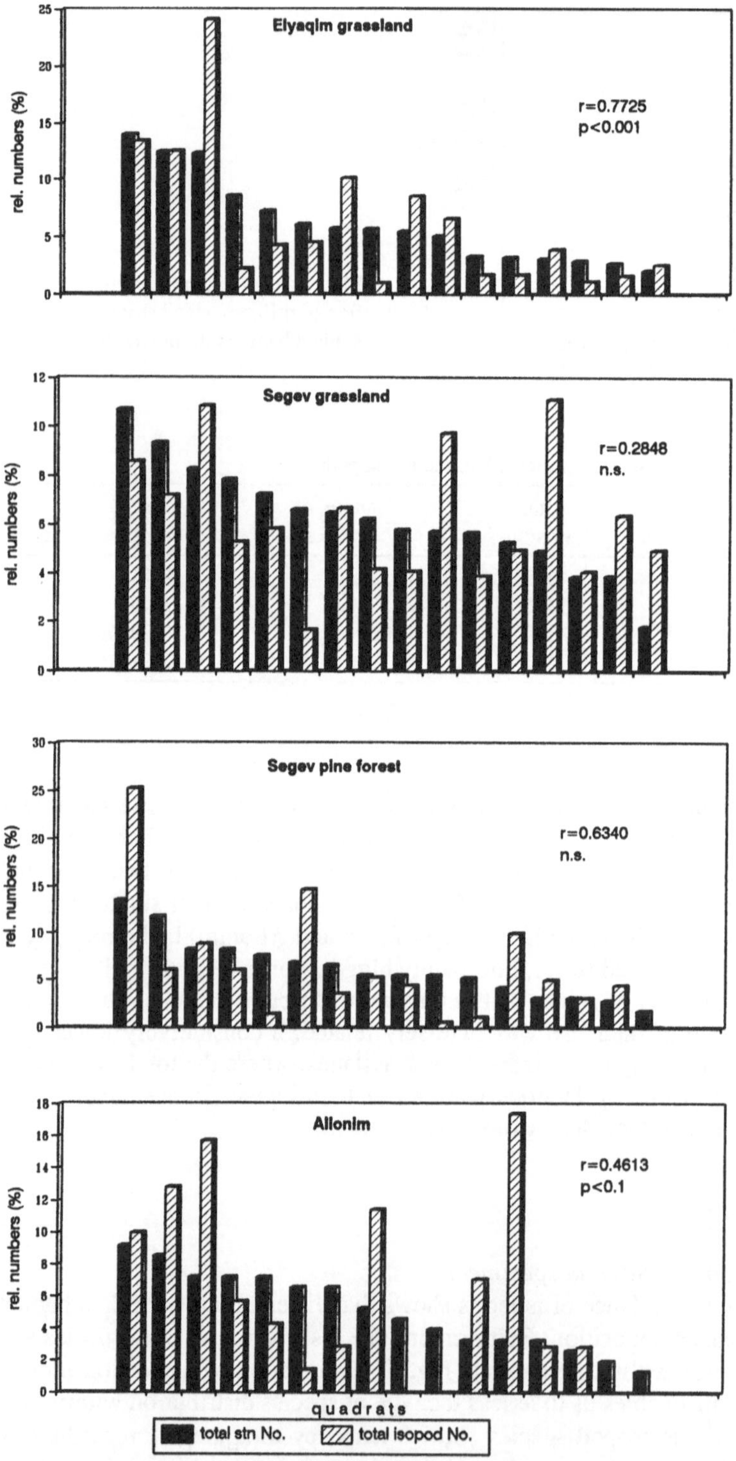

Figure 5. Distribution of stones and isopods, and their correlation in the 16 quadrats within each habitat. Quadrats are ranked by their stone numbers.

Table 3. Values of dispersion indices of isopods in the four habitats. <1 segregated; ~1 random; >1 aggregated dispersion type.

	Habitats Elyaqim	Segev-gr	Allonim	Segev-fo
total isopods	1	0.43	0.96	0.87
Porcellio pruinosis	0.78	0.51	1.27	1.51
Chaetophiloscia spp.	0.95	0.51	1.15	0.82
Armadillo officinalis			2.39	0.98
Sch. tiberianum	1.34		1.89	1.5
Porcellio ficulneus	1.6			
Armadillo sp. 'brown'	2.31	1.03		

3.2.2 *Isopod number-stone coverage correlation.*

Apparently, by the results of regression analysis, there is a strong correlation between the whole stone coverage (i.e. the total number of stones) and the number of isopod specimens in Elyaqim grassland, but only a slight one in Allonim. No significant correlation was found with the total stone number in both Segev habitats (Fig. 5).

Different sized stones have different importance for isopod specimens. In Elyaqim grassland there is a significant correlation between isopod number and all stone-classes, regardless of their size. In the other three habitats only medium and/or large sized stones are of significant attractiveness for isopods (Table 4).

3.3 *Minidistribution*

As the species occuring together under the same stone were registered, a chi-square test was applied for the association of species found under the same stone in all habitats (Table 5). In grassland habitats (Elyaqim, Segev-grassland) there appears to be a significant association between most of the species. In the woody habitats there was no significant association (Segev-forest), or it was occasionally significant (Allonim). Some species frequently shared their shelters (*P.pruinosus* and *Armadillo* sp.), while others (*S.tiberianum* and either *Chaetophiloscia* spp or *A.officinalis*) hardly ever shared stone microhabitats.

4 DISCUSSION

The Mediterranean region occupies most of the coastal plain and the mountainous region in Israel (Fig. 1a). Examples for the distribution of the six most abundant isopod species within this region are given in distribution maps (Fig. 1b-d).

From the recent publications of Schmalfuss (1988, 1991, 1992) it appears that some of the species discussed also in the present study have a West-Asiatic (*S.tiberianum, P.ficulneus*), or Eastern Mediterranean (*Chaetophiloscia* spp.) zoogeographical distribution. The occurence of *P.ficulneus* was proved only in the Mediterranean region of Israel (Schmalfuss 1992).

In previous studies, 16 isopod species were identified in the Mediterranean region

Table 4. Coefficients (r) and significance levels (p) of correlation between stone and isopod numbers. r = correlation coefficient. p = significance level.

No. of stones		Habitat Elyaqim	Segev-gr No. of isopods	Segev-fo	Allonim
total	r	0.77	0.28	0.31	0.46
	p	0.001	n.s	n.s.	0.1
small	r	0.75	0.16	-0.18	0.16
	p	0.001	n.s.	n.s.	n.s.
medium	r	0.72	0.62	0.55	0.41
	p	0.01	0.01	0.05	n.s
large	r	0.46	0.17	0.53	0.62
	p	0.1	n.s.	0.05	0.01

Table 5. Isopod species co-occurence in the four habitats (significance of Chi^2 test). The test was made for the most abundant species missing values indicate no common occurence. Key to the species: P.f. = *Porcellio ficulneus*, P.p. = *Porcellionides pruinosus*, C.spp. = *Chaetophiloscia* spp., A.sp. = *Armadillo officinalis*, S.t. = *Schizidium tiberianum.*

Species	Habitat Elyaqim	Segev-gr	Allonim	Segev-fo
P.f.-P.p.	< 0.01			
P.f.-C.spp.	< 0.01			
P.f.-A.sp.	< 0.01			
P.f.-S.t.	< 0.01			
P.p.-C.spp.	< 0.01	< 0.01	< 0.02	n.s.
P.p.-A.sp.	< 0.01			
P.p.-S.t.	< 0.01		n.s.	
C.spp.-A.sp.	n.s.	< 0.01		
C.spp.-S.t.	n.s.		n.s.	n.s.
A.sp.-S.t.	< 0.01			
A.o.-P.p.			< 0.01	n.s.
A.o.-C.spp.			< 0.01	n.s.
A.o.-S.t.			n.s.	n.s.

of Northern-Israel (Warburg et al. 1978). These included Allonim, where at that time 9 species were present. The most abundant of these at Allonim were *Chaetophiloscia* spp., *P.pruinosus*, while *A.officinalis* and *S.tiberianum* were subdominant. Isopod species diversity there, calculated by the same index, reached a higher value then was determined in the present study (1.05 vs 0.66). In that one year-long survey more species were found due to their seasonal phenology. Some species make their apparence during fall, whereas others appear during spring and even early summer. Such species could not be observed in the habitats sampled as the present study was conducted in winter and spring. Changes in population structure in the three most abundant species are described by Warburg et al. (1984), who showed that the main activity period of most isopod species was during February and March. The phenomenon seems to be particularly apparent in these xeric habitats. In the more arid regions a similar pattern was noticable (Kheirallah 1979, 1980).

All the habitats, other than Allonim, studied here were previously not mentioned. They did not vary in number of isopod species but in their species composition and abundance. Therefore it was apparent, that isopods are not evenly distributed among the various habitats within the Mediterranean region of Northern Israel. The segregation of species based on habitat level can be best seen in the case of the two adjacent Segev habitats, within the same locality. Although there is no physical border between the two sites, they differ in their species composition and have a low similarity index (0.17). This can be due to some kind of ecological isolation: among the abundant species, *A.officinalis* and *S.tiberianum* live only in the pine forest, while *Armadillo* sp. ('brown') was found only in the grassland (Table 1).

In a previous study a positive correlation between stone coverage and isopod abundance was also found (Warburg et al. 1978). Stones proved to be most important for isopods in large tract of open grassland (Elyaqim). No similar studies comparing the relationship between isopod abundance and stone coverage appear to have been made. We also found that isopod species were not evenly distributed within each of the four habitats. At the species level we found dispersion to be mostly clumped. The 'within habitat' distribution of the same species differed in the different habitats.

In Europe, the microdistribution of *Armadillidium vulgare* and *Trachelipus nodulosus* in temperate grasslands exhibited aggregated dispersion that was due to a microclimatically patchy environment (Hornung 1984, 1989, 1991a, 1991b). In England similar findings by Brereton (1957), described 'preference' of different microhabitats under the same wood. All 5 isopod species studied by him showed a definite 'preference' of a particular microhabitat. *Porcellio scaber* was hardly ever found in litter or in dead wood, but mostly at the lower level of tree trunks. Both *Philoscia muscorum* and *Trichoniscus* spp. showed 'preference' to leaf litter and dead wood, and were never found on the tree trunk. This pattern changed seasonally, and in summer more species (and specimens) were found in leaf litter and dead wood. Similarly, Davies & Sutton (1977) found that four isopod species behaved differently in their aggregation response in different microsites within the same habitat. The fact that isopods tend to aggregate in certain microhabitats was previously described by Allee (1926) and Cole (1946). Cole found that isopods 'prefered' to aggregate in groups of 3-5 animals, under boards that covered the ground and thus were moist underneath. This aggregation pattern changed seasonally. In a dune area, *A.vulgare* showed varying seasonal abundance under stones (Davis & Sutton 1977). A similar observation was reported for *Periscyphis granai* in a xeric habitat from Saudi Arabia (Kheirallah 1979).

This is the first time that an analysis has been made for species association in terrestrial isopods. The species associations observed in the present study can be only due to cohabitation of species which share the same shelter. In spite of the fact that there are numerous studies on the dynamic changes in isopod population structure (Dangerfield 1987; Davis 1984; Sunderland et al. 1976; Sutton 1968), there is no information regarding the micro- or minidistribution of isopods within the habitat.

5 CONCLUSIONS

Based on the present study we can state that the zoogeographical scale used by taxonomists can be only a starting point to ecological field works. The zoogeographical distribution even within a comparatively small region is not intended to provide information on the local distribution within various types of habitats. It cannot reflect the real distribution of a species within a region. Species are segregated on habitat level and not evenly distributed among those (macrodistribution).

In the present study, the microdistribution of isopods within a habitat was related to stone coverage. Thus, in open habitats stone coverage is important for isopods in the Mediterranean region probably because of the harsh microclimate there. Fewer stones in a habitat imply a higher rate of their occupation by isopods who use them as shelters. The larger stones seem to shelter greater numbers of specimens.

Although in some cases we can prove statistically significant species co-occurence under stones (minidistribution), it may be due only to the common use of the same resource as shelter in particular in grasslands and less so in woody habitats.

ACKNOWLEDGEMENTS

Cordially thanks are due to an anonymous reviewer for his helpful comments improving the manuscript. The support of travel grant by the Hungarian Academy of Sciences (OTKA I/3-1607) to the senior author is gratefully aknowledged. One of us (EH) gratefully acknowledges the support by Technion during her stay in Israel.

6 REFERENCES

Allee, W.C. 1926. Studies in animal aggregations. Causes and effects of bunching in land isopods. *J.Exp.Zool.* 45:255-277.

Beyer, R. 1964 Faunistisch-ökologische Untersuchungen an Landisopoden in Mitteldeutschland. *Zool.Jb.Syst.* 91:341-402.

Brereton, J.Le G. 1957. The distribution of woodland isopods. *Oikos* 8:85-106.

Chelazzi, G. & F.Ferrara 1978. Researches on the coast of Somalia. The shore and the dune of Sar Uanle. 19. Zonation and activity of terrestrial isopods (Oniscoidea). *Monitore Zool. Ital. (N.S.) Suppl.* 11:189-219.

Cole, L.C. 1946. A study of the cryptozoa of an Illinois woodland. *Ecol.Monogr.* 16:49-86.

Dangerfield, J.M. 1987. Variations in the life history and population parameters of *Armadillidium vulgare* (Latreille). Unpublished Ph.D.Thesis, UEA, UK, 252pp

Davis, R.C. 1984. Effects of weather and habitat structure on the population dynamics of three species of woodlice (Crustacea: Oniscoidea) in dune grassland. *Oikos* 42:387-395.

Davis, R.C. and S.L.Sutton 1977. Spatial distribution and niche separation of woodlice and millipeds in a dune grassland ecosystem. *Ecol.Bull.(Stockholm)* 25:45-55.

Herold, W. 1930. Beiträge zur Verbreitung und Ökologie der Landisopoden des Ostbalticums. *Z.Morphol.Ökol.Tiere* 18:474-535.

Hornung, E. 1984. Characteristics of the population of an isopoda species (*Trachelipus nodulosus* C.L.Koch) at sandy soil grassland. *Acta Biol.Szeged.* 30:153-158.

Hornung, E. 1989. Population dynamics and spatial distribution of *Trachelipus nodulosus* (C.L.Koch, 1838) (Crustacea Isopoda) in a sandy grassland. *Monitore Zool. Ital. (N.S.) Monogr.* 4:399-409.

Hornung, E. 1991. Isopod distribution in a heterogeneous grassland habitat. In P.Juchault & J.P.Mocquard (eds.), *Third Symp. on the Biology of Terrestrial Isopods.* Poitiers, 73-79.

Hornung, E. 1992. Comparision of different grassland types based on isopod communities. *Proc.of the 4th ECE/XIII.SIEEC, Gödöllô* 1991, 741-746.

Kheirallah,A.M. 1979. The population dynamics of *Periscyphis granai* (Isopoda: Oniscoidea) in the western highlands of Saudi Arabia. *J.Arid Environ.* 2:329-337.

Kheirallah, A.M. 1980. Aspects of the distribution and community structure. *J.Arid Environ.* 3: 69-74.

Miller, M.A. 1938. Comparative ecological studies on the terrestrial isopod Crustacea of the San Francisco Bay region. *Univ.Calif.Publ.Zool.* 43:113-141.

Schmalfuss, H. 1988. The terrestrial isopod genus *Schizidium* in Western Asia (Oniscoidea: Armadillidiidae). *Stuttgarte Beitr. Naturk. Ser.A*, Nr.423:1-22.

Schmalfuss, H. 1991. The terrestrial isopod genus *Chaetophiloscia* in Western Asia (Oniscidea: Philosciidae). *Stuttgarter Beitr. Naturk.* (A)463:1-9.

Schmalfuss, H. 1992. The terrestrial isopod genus *Porcellio* in Western Asia (Oniscidea: Porcellionidae). *Stuttgarter Beitr. Naturk.* (A)475:1-45.

Sunderland, K.D., M.Hassall & S.L.Sutton 1976. The population dynamics of *Philoscia muscorum* (Crustacea, Oniscoidea) in a dune grassland ecosystem. *J.Anim.Ecol.* 45:487-506.

Sutton, S.L. 1968. The population dynamics of *Trichoniscus pusillus* and *Phyloscia muscorum* (Crustacea, Oniscoidea) in limestone grassland. *J.Anim.Ecol.* 37:425-444.

Warburg, M.R., D.Rankevich & K.Chasanmus 1978. Isopod species diversity and community structure in mesic and xeric habitats of the Mediterranean Region. *J.Arid Environ.* 1:157-163.

Warburg, M.R., K.E.Linsenmair & K.Bercovitz 1984. The effect of climate on the distribution and abundance of isopods. *Symp.Zool.Soc.Lond.* 53:339-367.

Diversity and spatial community stucture of terrestrial isopods (Isopoda, Oniscidea) in a mosaic of plant assemblages

Katalin Szlávecz
Soil Zoology Group, Dept. of Systematic Zoology and Ecology, Eötvös University, Budapest, Hungary, and Dept. of Geography and Environmental Engineering, The Johns Hopkins University, Baltimore, USA

ABSTRACT

The terrestrial isopod fauna of the Bátorliget Nature Reserve, a small area of exceptionally high biodiversity, was studied. A total of eighty pitfall traps were placed in different plant assemblages, such as wet gallery forests, drier oak forests, meadows, and birch swamp. The pitfall traps caught six species of isopods. Species composition and abundance varied markedly among sites, although isopod communities did not correspond to plant community classifications. The relatively undisturbed oak-elm-ash gallery forest and the birch swamp are the habitats for *Hyloniscus transsilvanicus*, a relic isopod from the post-glacial period. *Armadillidium vulgare* was the most abundant, *Haplophthalmus hungaricus* the least abundant species in the nature reserve. Total isopod abundance was negatively correlated, whereas diversity and evenness were positively correlated with habitat moisture conditions. The highest number of isopods were caught in spring and summer. Landscape level patchiness seems to be crucial in maintaining high biodiversity in this area. Conservation efforts, therefore, should focus on protecting this habitat mosaic along with preventing the area from further drying.

1 INTRODUCTION

Nature reserves are of great importance in conservation biology despite all debates about them. The Bátorliget Nature Reserve is one of the most remarkable protected areas in Hungary. Its outstanding floristic and faunistic values have long attracted botanists and zoologists. As a result of the intensive research, that started about 70 years ago, Bátorliget became one of the best explored sites of the world. Its biodiversity is remarkable: about 20% of the entire known Hungarian flora and fauna lives on an area less than one km^2.

Parts of the area became protected in 1951, one among the first nature reserves in Hungary. The botanical and zoological surveys during the first half of the century resulted in a monograph (Székessy 1953), which attracted international attention.

In the 1980s, a new project was initiated and coordinated by the Hungarian Natural History Museum. The objectives of this survey were: 1) To compare the species composition of the area to the one published forty years ago, 2) To detect any changes since the area became protected, 3) To give a more complete zoogeographical evaluation, 4) To make recommendations for future conservation management. The results of this research are summarized in a two volume book (Mahunka 1991a).

We studied the macroarthropod fauna of Bátorliget. Species lists and zoogeographical evaluation are given in Allspach & Szlávecz (1991), Loksa (1991a,b), Korsós (1991).

The area is unique, because it preserved the prehistoric landscape of the Hungarian Great Plains now consisting mostly of arable fields and dry steppe. Fragments of once huge marshlands and different forest types still exist here. Human influence (drainage, clearcutting, subsequent reforestation) further diversified the landscape resulting in a mosaic of various plant assemblages. In addition to the above mentioned objectives, we were also interested in the similarities and differences among the soil invertebrate communities of these various vegetation types. Soil macroarthropod communities in general are compared in Szlávecz & Loksa (1991). A more detailed study was carried out on the spatial distribution and abundance of terrestrial isopods. The present paper reports the results of this analysis.

2 MATERIAL AND METHODS

The Bátorliget Nature Reserves (BNR thereafter) is located in the northeastern part of Hungary, in an area called Nyirség (Fig. 1). It consists of three separate areas: a mire ('Bátorligeti-láp'), a forest ('Fényi erdö') and a pasture ('Bátorligeti legelö'). The area of the mire reserve, where the present study was conducted, is very small, about 60 ha. The majority of soil types in the mire reserve are of alluvial origin with the exception of the dunes which are covered with wind blown humic sandy soils (Rajkai 1991). Out of the many vegetation types on the study area we chose eight sites (denoted with letters A-H thereafter) for our survey (Fig. 1).

A and B: Oak-elm-ash gallery forests. Temporal flooding occurs on these sites. The forests are rich in geophytes. Site B is an old seminatural stand, whereas site A was reforested forty years ago, when the area became protected. The characteristic soil type is meadow soil.

C and D: Sandy pedunculate oak-silver lime forests. These stands grow on sand dunes and are relatively dry. Both sites used to be arable fields until 1951, when they were reforested. Grasses are characteristic to the understory. The soil type here is 'kovárvány' humic sandy soil.

E, F and G: Meadows. All three are wet meadows, although sites E and F will be later referred to as 'drier' meadows. This simply means that they are covered with water for shorter periods, than site G, which could also be called a wetland. The latter is covered with dense moss layer. E and G are regularly mowed, whereas on F trees and shrubs are regenerating. The soil types on E, F, G are peaty meadow soil, meadow soil and drying peat soil, respectively.

H: Birch swamp. This is a natural swamp with some drier patches, on which predominantly aspen trees grow.

More detailed description of the flora of Bátorliget is given in Standovár et al. (1991).

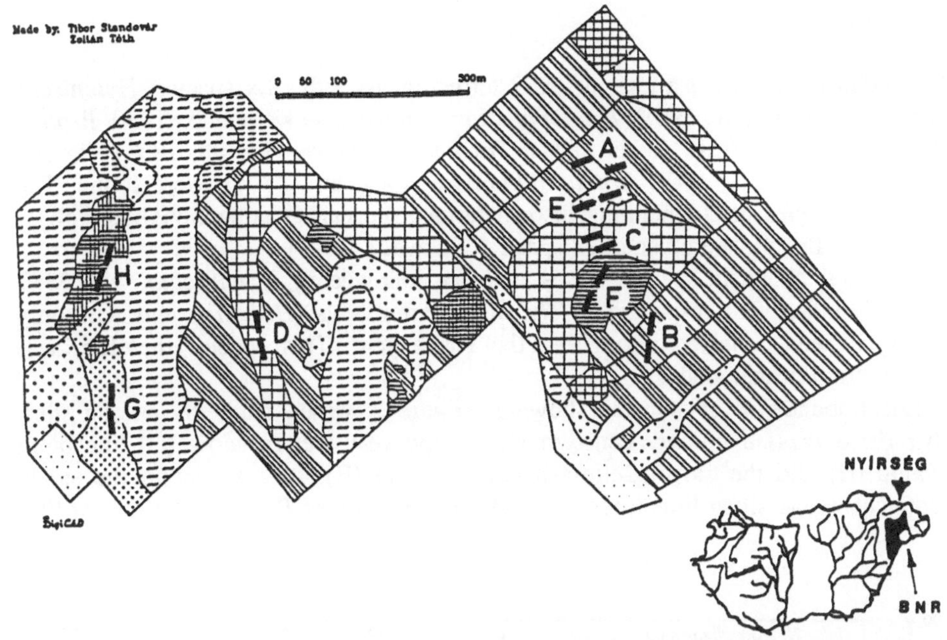

Figure 1. Location of the Bátorliget Nature Reserves (BNR) in Hungary and vegetation map of the mire reserve with sites of collection (A-H). Each line represents five pitfall traps in a row. A, B) Oak-elm-ash gallery forests, C, D) Sandy pedunculate oak-silver lime forests, E, F, G) Meadows, H) Birch swamp. Description of the plant communities are given in Material and methods section in text.

In November 1989 ten pitfall traps, filled with ethylene glycol, were placed on each site. The pitfall traps were emptied four times: on 20/3/1990, 13/6/1990, 18/7/1990 and 19/10/1990. The material was sorted and stored in the laboratory, and then each isopod was identified.

Diversity (*H*) was calculated using the Brillouin-index for finite collections:

$$H = \frac{1}{N} \ln \frac{N!}{N_1! N_2! \ldots N_i! \ldots N_s!}$$

where N_i is the number of *i*-th species, *s* is the total number of species. Evenness (*J*) was calculated as follows:

$$J = \frac{H}{H_{max}}$$

$$H_{max} = \frac{1}{N} \ln \frac{N!}{(X!)^{s-r} (Y!)^{r}}$$

where $X = (N/s)$, the integer part of N/s, r is the remainder, and $Y = X + 1$ (Pielou 1975).

3 RESULTS

The pitfall traps caught a total of 6972 terrestrial isopods. Six species, *Hyloniscus transsilvanicus* (Verh., 1901), *Haplophthalmus hungaricus* Kesselyák, 1930, *Protracheoniscus politus* (C.L. Koch, 1841), *Trachelipus rathkei* (Brandt, 1833), *Porcellium collicola* Verh., 1907, *Armadillidium vulgare* (Latr., 1804), were found. The complete species list (containing also material from earlier collections) as well as faunistic evaluation are given in Allspach & Szlávecz (1991).

The most abundant species was *A. vulgare* followed by *P. collicola*, with 3387 and 2259 individuals respectively. The other extreme was *H. hungaricus* with only two specimens collected throughout the entire year. This species therefore was omitted from further analysis.

Local abundance of isopods, shown in Figure 2, varied markedly among sites. *Hyloniscus transsilvanicus* occurs mainly in the very wet meadow (G), the birch swamp (H) and the old oak-elm-ash gallery stand (B). It is absent in the sandy pedunculate oak-silver lime forests (C, D). The majority of *Protracheoniscus politus*

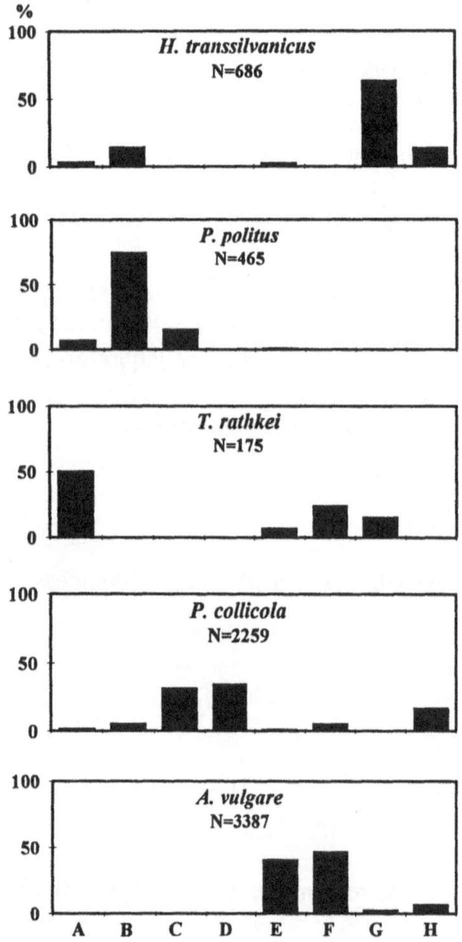

Figure 2. Distribution of the isopods in the different vegetation types. Bars represent the percentage of individuals caught throughout the entire sampling period (*N*). A, B) Oak-elm-ash gallery forests, C, D) Sandy pedunculate oak-silver lime forests, E, F, G) Meadows, H) Birch swamp.

occurs also in the seminatural gallery forest (B). The abundance of the ubiquitous *Trachelipus rathkei* seems to be generally low in the whole area. About half of the isopods were caught in the 40 year old planted oak-elm-ash gallery forest (A). *Porcellium collicola* was much more evenly distributed among habitats. It was the dominant species, almost the exclusive inhabitant in the two sandy oak forests (C, D). *Armadillidium vulgare*, on the other hand, was found in the open habitats (E, F, G), and in the birch swamp (H). Very few (a total of thirty) individuals were caught in the four forests.

Total isopod abundance was negatively correlated with habitat moisture conditions (Table 1). [Ranking is based upon soil moisture conditions (Rajkai 1991), microclimate measurements (Soó 1953) and the length of the period a given site is covered with water]. On the species level the abundance of *H. transsilvanicus* was clearly highest on the most humid sites (Fig. 3), whereas *P. collicola* showed an opposite trend. There was no correlation between these two attributes for the other species (Table 1).

Isopod abundance fluctuated not only spatially but temporally, as well (Fig. 4). All species showed a similar pattern: the highest number of individuals were caught in spring and summer, whereas autumn and winter were much more inactive periods. Specific differences could still be detected: e.g. the peak of activity for *P. collicola* was spring, followed by a decline in summer, whereas *T. rathkei* showed the opposite trend.

Table 1. Correlation between habitat moisture conditions and isopod abundance, species diversity, and evenness for species collected in pitfall traps (*Haplophthalmus hungaricus* not included due to insufficent numbers collected). r = Spearman rank correlation coefficient; n.s. = not significant.

	r	P
Hyloniscus transsilvanicus	0.92	0.01
Protracheoniscus politus	–0.31	n.s.
Trachelipus rathkei	0.38	n.s.
Porcellium collicola	–0.66	0.05
Armadillidium vulgare	0.31	n.s.
total number of isopods	–0.67	0.05
Diversity (h)	0.76	0.05
Evenness (J)	0.74	0.05

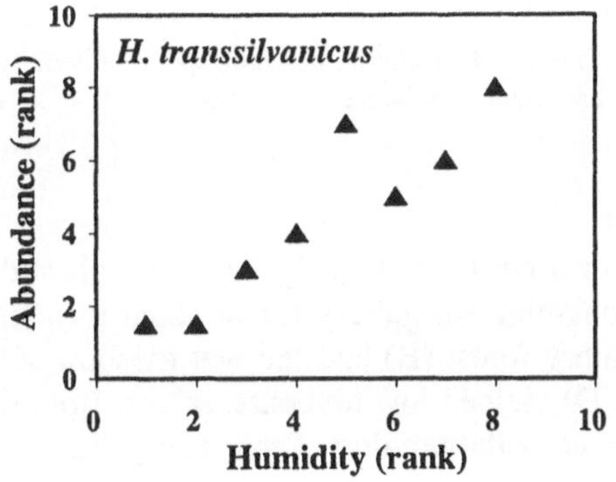

Figure 3. Correlation between abundance of *H. transsilvanicus* and habitat moisture conditions.

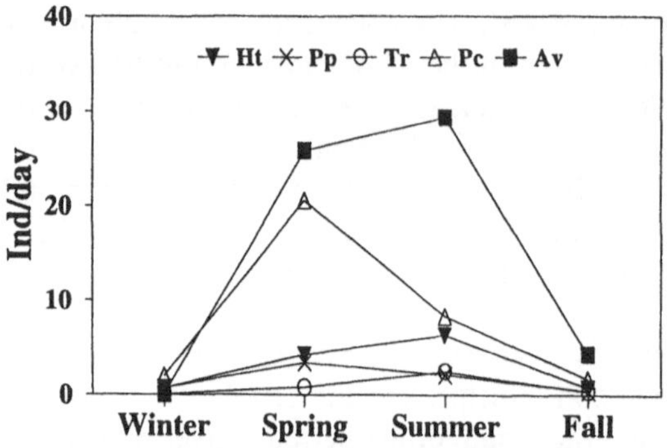

Figure 4. Seasonal fluctuations in isopod activity. Values expressed as number of individuals caught per day. Data for the different habitats are pooled. Ht = *Hyloniscus transsilvanicus*, Pp = *Protracheoniscus politus*, Tr = *Trachelipus rathkei*, Pc = *Porcellium collicola*, Av = *Armadillidium vulgare*.

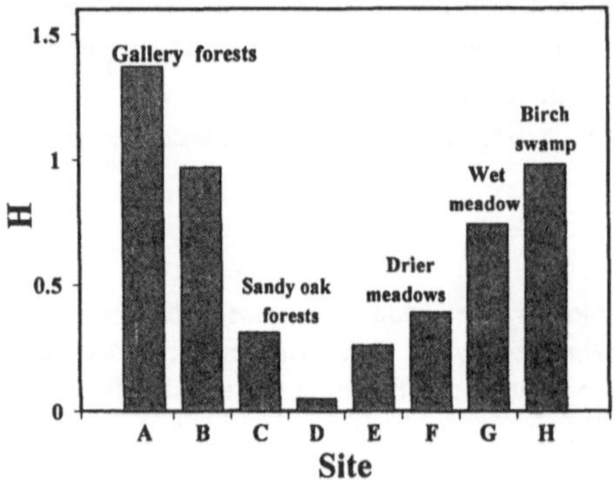

Figure 5. Diversity (*H*) of isopods in the different plant assemblages.

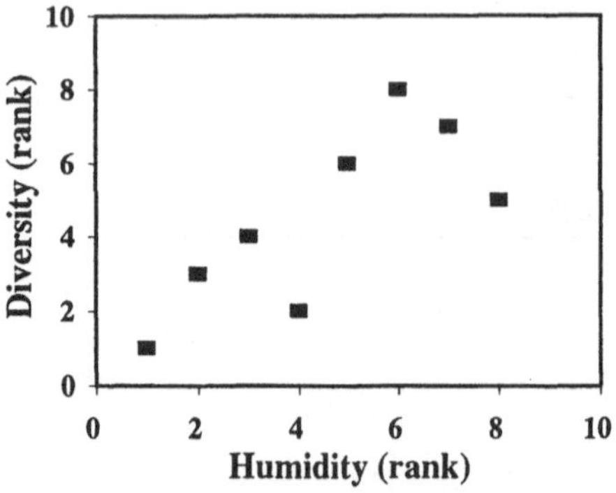

Figure 6. Correlation between species diversity and habitat moisture conditions.

Diversity of isopods among habitats also varied markedly (Fig. 5). The highest diversity index was obtained in the young oak-elm-ash gallery forest stand (A), followed by the birch swamp (H), the old gallery forest (B) and the wet meadow (G). The two sandy pedunculate oak forests (C, D) yielded low diversity values. Both diversity and evenness were positively correlated with humidity (Table 1, Fig. 6).

4 DISCUSSION

Some of the observed patterns in the local distribution of the different isopod species are similar to those obtained by Tomescu et al. (1979), who studied the arthropod fauna on a southern slope in the mountains of Transylvania, Romania. In both areas *Hyloniscus* species were restricted to the wettest sites. *Trachelipus rathkei* was represented with very low numbers. *Protracheoniscus politus* occurred in forest communities only, similarly to the findings of Flasarová (1986) in the Little Carpathians. *Porcellium collicola* was a common species (in the Romanian study also the most abundant one), occurring in all vegetation types. This was the result of another survey in the Pilis Biosphere Reserve, Hungary, as well (Szlávecz 1988). On the other hand, there are differences in the local distribution of *Armadillidium vulgare* populations. In Bátorliget almost ninety percent of all individuals occurred in the two drier meadows (E, F). The areas of these are extremely small: the smaller (E) is about 60 m in diameter (Fig. 1). Since this isopod species does not seem to occur in the surrounding gallery forests (A, B), the populations are concentrated on very restricted areas (small probably even on the scale of an isopod population). In the Romanian study site, however, *A. vulgare* is present in the woodlands only. One of the major differences between the two study sites is that the open habitats in Romania are very dry (perhaps too dry even for such a drought tolerant species like *A. vulgare*), whereas the mire reserve in Bátorliget is generally a wet area with the patches being less or more humid. It is the absence of *A. vulgare* in the forest habitats of the mire reserve that is surprising. In another protected area of the Great Hungarian Plains, both open grasslands and forests had *A. vulgare* populations, but more animals were caught in the latter (Loksa 1973; Szlávecz 1991).

Distribution of *H. transsilvanicus* is clearly influenced by moisture conditions (Fig. 3). Flasarová (1980) collected this species in the Little Carpathian Mountains of Slovakia in a beech-oak-alder forest along a creek. It was also the most abundant species in a moist alder forest in Romania (Radu & Tomescu 1976). Not much is known about the habitat selection of *H. transsilvanicus*, but since it belongs to a genus known to be extremely sensitive to moisture conditions (e.g. Gruner 1966), it is not surprising, that the distribution of this species is also limited by humidity.

It has been shown that climate affects isopod acivity (e.g. Warburg et al. 1984, Hornung 1989). In Central Europe, isopods are generally least active during winter. Extremely hot and dry summer can also result in decreased activity especially in open areas. This is not the case in Bátorliget, where the microclimate is generally cooler than on the surrounding arable fields and steppe vegetation (Soó 1953). The reason for this is the proximity of the cool ground water to the surface. This keeps the soil moist and cool, thus the air above ground stays also cool and humid. The surrounding dunes and forests protect this air from the wind, resulting in a cool, moist microclimate even in summer. With less daily and seasonal temperature and moisture fluctuations, isopods stay active throughout the whole growing period. It is important to point out, however, that changes in the numbers of animals in pitfall traps do not necessarily reflect changes in activity. An increase can be the result of greater activity and/or appearance of the new generation. Here, with the exception of *A. vulgare*, juveniles were represented in very low percentages in the pitfall traps, therefore the observed seasonal changes in abundance correspond to changes in surface activity.

Isopod diversity was high in the two gallery forests (A, B), the wetland (G), and the swamp (H), and low on the remaining sites (Fig. 5). The other major saprophagous arthropod group, the Diplopoda, showed the same pattern. The predatory soil macroarthropods (Chilopoda, Opiliones and Araneae) had high species diversity in the most ancient, least disturbed (by humans) habitats, but the overall pattern was different. Spiders, for instance, had high diversity on all sites, whereas harvestmen were absent on the wetland (Szlávecz & Loksa 1991).

Disturbance theory is one of the most dynamic fields in recent community ecology. It has also practical significance in conservation and agriculture (Begon et al. 1990). With a number of hypotheses available on the role of disturbance in structuring communities (e.g. Connell 1978; Yodzis 1986), one might wonder about the correlation between diversity and degree of disturbance in BNR. In my opinion, it is inapproriate to carry out such an exercise on the Bátorliget data. The landscape is very diverse, each plant assemblage has its own history. The nature and intensity of disturbances on the different vegetation types differ so much, that it is not possible to establish even a ranking order among them. Also, different disturbances (e.g. clear-cutting, earlier agricultural practice, mowing, flooding) operate on very different time scales, which further complicates the situation. The reason, why the most ancient sites have high isopod diversity, is that these fragmented plant communities preserved the vegetation once so characteristic to the prehistoric landscape of the whole Nyirség. This, together with the favourable microclimatic (primarily moisture) conditions ensures the survival of the rare and relic isopods. At the same time more common species are also present on these sites.

Although the distribution of isopods is not uniform, their communities do not necessarily correspond to plant community classifications. Cluster analysis revealed, that soil macroarthropod communities of the oak-elm-ash gallery forests (old and reforested stands, B and A, respectively) and the birch swamp (H) are the most similar to one another (Szlávecz & Loksa 1991). The two drier meadows (E, F) and the two sandy pedunculate oak-silver lime forests (C, D) formed two other groups. These drier habitats have also very low isopod diversity but high total abundance (Table 1). Sites C and D, the two sandy oak forests are inhabited by *P. collicola* almost exclusively, whereas on the two meadows (E, F) *A. vulgare* is highly dominant. Although the latter two sites are different from each other, in that meadow F, unlike E, is not mowed, therefore trees and shrubs started to regenerate on it, their isopod species composition and abundance are very similar.

H. transsilvanicus is undoubtedly the faunistically most exciting isopod in Bátorliget (Méhely 1929; Székessy 1953; Allspach & Szlávecz 1991). Efforts to protect this relic species should focus on preserving its habitats, the gallery forests, the birch swamp and the very wet meadow (B, G, H, respectively). Reforestation of the old fields, that occurred forty years ago, seems to be effective in recreating favourable conditions for isopods: the young stand (A) was invaded by *H. transsilvanicus*, and *P. politus*, another rare isopod in Hungary. The distribution of these species and a third rare isopod, *Haplophthalmus hungaricus*, is restricted in moist habitats (in case of *P. politus* in humid forests). Slow drying of the area, indicated by long term changes in plant species composition (Standovár et al. 1991) threatens survival of these isopod populations in BNR. Effective manipulation of the water table is one way this problem can be managed (Simon 1991).

The sandy oak forests do not harbour isopods of faunistic importance. This is not the case for other soil animal groups. Many species, with the only known locality in Bátorliget, occur in this vegetation type. Zoogeographically these curiosities are of mediterranean rather than boreal origin. This is the case for instance for oribatid mites (Mahunka 1991b) and Collembola (Loksa 1991c). Therefore on a landscape level the heterogeneity of the habitat, brought about by natural and human disturbances, ensures the high diversity of the whole nature reserve. This means that long term efforts to preserve the exceptional biodiversity of Bátorliget should focus on maintaining the mosaic nature of the habitat. Also, to detect any future changes, a well organized monitoring scheme is necessary.

ACKNOWLEDGEMENT

I am most grateful to my late professor, Dr. I. Loksa for his continuous encouragement and help throughout this study. Assistance by M. Seidl and P. Tóth in the field and in the laboratory is also greatly appreciated.

REFERENCES

Allspach, A. & K. Szlávecz 1991. The terrestrial isopod (Isopoda: Oniscidea) fauna of the Bátorliget Nature Reserves. In S. Mahunka, (ed.), *The Bátorliget Nature Reserves – After Forty Years.* Studia Naturalia 1, Budapest: Hung. Nat. Hist. Mus. Publ.

Begon, M., J.L. Harper & C.R. Towsend 1990. *Ecology: Individuals, Populations and Communities.* Boston: Blackwell.

Connell, J.H. 1978. Diversity in tropical rainforests and coral reefs. *Science* 199: 1302-1310.

Flasarová, M. 1980. *Hyloniscus transsilvanicus* (Verhoef 1901) im Gebirge Malé Karpaty in der Westslowakei. *Faun. Abh. Mus. Tierk. Dresden* 7: 273-278.

Flasarová, M. 1986. Isopoda (Asellota, Oniscoidea) of the Little Carpathians. In J. Nosek (ed.), *The soil fauna of the Little Carpathians.* Bratislava: Vydavatelstvo SAV.

Gruner, H. 1966. Isopoda. In M. Dahl & F. Peus, (eds.), *Die Tierwelt Deutschlands.* Jena: VEB Gustav Fischer Verlag.

Hornung, E. 1989. Population dynamics and spatial distribution of *Trachelipus nodulosus* (C.L. Koch, 1838) (Crustacea, Isopoda) in a sandy grassland. *Monitore Zool. Ital. (N.S.) Monogr.* 4: 399-409.

Korsós, Z. 1991. The centipede and millipede fauna of the Bátorliget Nature Reserves (Chilopoda and Diplopoda). In S. Mahunka (ed.), *The Bátorliget Nature Reserves – After Forty Years.* Studia Naturalia 1, Budapest: Hung. Nat. Hist. Mus. Publ.

Loksa, I. 1973. Bodenzoologische Untersuchungen in den Alkali-Waldsteppen von Margita, Ungarn. 1. Untersuchungen der Arthropoden-Makrofauna, nebst Bemerkungen über die Oniscoidea-Arten. *Opusc. Zool. Budapest* 11:79-93.

Loksa, I. 1991a. The harvestmen (Opiliones) fauna of the Bátorliget Nature Reserves (NE Hungary). In S. Mahunka (ed.), *The Bátorliget Nature Reserves – After Forty Years.* Studia Naturalia 1, Budapest: Hung. Nat. Hist. Mus. Publ.

Loksa, I. 1991b. The spider (Araneae) fauna of the Bátorliget Nature Reserves. In S. Mahunka (ed.), *The Bátorliget Nature Reserves – After Forty Years.* Studia Naturalia 1, Budapest: Hung. Nat. Hist. Mus. Publ.

Loksa, I. 1991c. The Collembola fauna of the Bátorliget Nature Reserves. In S. Mahunka (ed.), *The Bátorliget Nature Reserves – After Forty Years.* Studia Naturalia 1, Budapest: Hung. Nat. Hist. Mus. Publ.

Mahunka, S. (ed.) 1991a. *The Bátorliget Nature Reserves – After Forty Years*. Studia Naturalia 1. Budapest: Hung. Nat. Hist. Mus. Publ.

Mahunka, S. 1991b: The oribatid (Acari: Oribatida) fauna of the Bátorliget Nature Conservation Areas. In S. Mahunka (ed.), *The Bátorliget Nature Reserves – After Forty Years*. Studia Naturalia 1, Budapest: Hung. Nat. Hist. Mus. Publ.

Méhely, L. 1929. Species generis *Hyloniscus. Studia Zool. Budapest* 1: 1-75.

Pielou, E.C. 1975. *Ecological Diversity*. New York: Wiley.

Radu, V. & N. Tomescu 1976. Quantitativ-ökologische Untersuchungen an Landisopoden. *Pedobiologia* 16: 36-43.

Rajkai, K. 1991. The soils of the 'Bátorligeti-Láp' Nature Reserve. In S. Mahunka (ed.), *The Bátorliget Nature Reserves – After Forty Years*. Studia Naturalia 1, Budapest: Hung. Nat. Hist. Mus. Publ.

Soó, R. 1953. A növénytakaró (The plant cover). In V. Székessy (ed.), *Bátorliget élövilága (The Natural History of Bátorliget)*. Budapest: Akadémiai Kiadó.

Simon, T. 1991. Nature conservation values of the Bátorliget area. In S. Mahunka (ed.), *The Bátorliget Nature Reserves – After Forty Years*. Studia Naturalia 1, Budapest: Hung. Nat. Hist. Mus. Publ.

Standovár, T., Z. Tóth & T. Simon 1991. Vegetation of the Bátorliget Mire Reserve. In Mahunka, S.(ed.), *The Bátorliget Nature Reserves – After Forty Years*. Studia Naturalia 1, Budapest: Hung. Nat. Hist. Mus. Publ.

Székessy, V. 1953. *Bátorliget élövilága (The Natural History of Bátorliget)*. Budapest: Akadémiai Kiadó.

Szlávecz, K. 1988. The isopod fauna of the Pilis Biosphere Reserve. I. Basaharc loess mine. *Opusc. Zool. Budapest* 23: 189-195.

Szlávecz, K. 1991. The terrestrial isopod fauna of the Hortobágy National Park. *Miscnea zool. hung.* 6: 61-66.

Szlávecz, K. & I. Loksa 1991. Diversity of soil arthropods in the Bátorliget Nature Reserve, Hungary. In L. Zombori & L. Pergovits (eds.), *Proc. Fourth Eur. Cong. Entomology, Gödöllö, Hungary*, pp. 801-807. Budapest: Hung. Nat. Hist. Mus. Press.

Tomescu, N., E.A. Schneider & I. Weiss 1979. Untersuchungen über die Arthropodenfauna xerothermer Standorte im südsiebenbürgischen Hügelland. IX. Die Isopoden eines Südhanges im Hügelland Südsiebenbürgens. *Muzeul Brukenthal, Studii si Communicari, St. nat.* 23: 275-286.

Warburg, M.R., K.E. Linsenmair & K. Berkowitz 1984. The effect of climate on the distribution and abundance of isopods. *Symp. zool. Soc. London* 53: 339-367.

Yodzis, P. 1986. Competition, mortality and community structure. In Diamond, J. and T.J. Case (eds.), *Community Ecology*. New York: Harper & Row.

Effects of temperature and photoperiod on the breeding patterns of two isopod species

M.R.Warburg & Dorit Weinstein
Department of Biology, Technion, Haifa, Israel

ABSTRACT

Temperature and photophase effects on reproductive patterns were studied experimentally in two oniscid isopod species: *Armadillo officinalis* (Armadillidae), an iteroparous species and *Schizidium tiberianum* (Armadillidiidae), a semelparous species. The photophase conditions ranged from total darkness to 10 hr, 14 hr and continuous light at both 17°C and 23°C. These were compared with normal reproductive pattern in the field. All experimental conditions disrupt normal reproductive patterns in both species by affecting the time tables of oögenesis, as well as the numbers of oöcytes, eggs and mancas. Consequently, oögenesis was shorter in both species resulting in early manca release, and oösorption appeared responsible for a decline in both marsupial egg and manca numbers. In *Armadillo* no marsupial formation took place at low temperature regardless of the photophase.

1 INTRODUCTION

The effect of both temperature and photoperiod on the breeding patterns of isopods has been studied in several species (see review in Warburg 1987, 1994; Souty-Grosset et al. 1988; Mocquard et al. 1989). The isopod species studied so far have been mainly from mesic habitats. Only recently has this aspect been studied in a xeric isopod species, *Porcellio ficulneus* (Hornung & Warburg 1993, 1994).

Whereas increased photoperiod had great significance in initiating reproduction in isopods, increased temperature accelerated it. In all these studies either the formation of marsupium or the release of juveniles were the main criteria for evaluating the effects of environmental factors on reproduction. In none of the studies was the effect of altered temperature and photoperiod evaluated by direct observations of the changes in ovarian or oöcyte dimensions, or of oöcyte numbers during oögenesis.

In the present study we selected two isopod species, both of which inhabit the Mediterranean region. They are an iteroparous species, *Armadillo officinalis* (Armadillidae) breeding in fall, and a semelparous species, *Schizidium tiberianum* (Armadillidiidae) that breeds in spring. Both were seasonally abundant and readily obtainable. Their reproductive patterns are described in Warburg & Cohen (1991, 1992a).

Xeric habitats within the Mediterranean region are characterized by short winter and long, hot periods in summer and fall. We therefore expected temperature and light to affect breeding patterns in the xeric-inhabiting isopods which are found only in certain seasons, differently than they do in the mesic- inhabiting species, which can be found during the whole year (Warburg et al. 1984). Moreover, we expected to find differences in the reproductive patterns of these two species as they may respond in different ways to changes in either temperature or photophase regimes. The reason for this being their different reproductive seasons. One, breeding in spring with oögenesis taking place during winter, whereas the other, a fall breeder having a summer oögenesis.

The following is an account of temperature and photophase effects on ovarian and oöcyte growth and development in these two isopod species.

2 MATERIALS AND METHODS

Animals were collected from grassland, oak woodland and pine forest habitats in the Lower Galil Mts. and on Mt. Carmel in northern Israel. Collecting took place once or twice monthly during the main activity season of *S. tiberianum* (December-April), and throughout most of the year for *A. officinalis*.

Freshly collected adult females were dissected regularly, others were placed individually in jars (with an adult male), and kept under experimental conditions for future periodical dissections and observations. The procedure involved placing the isopods under 7 different experimental conditions:

1. Continuous light (24 – L) at 25°C in a constant temperature room (± 1°C).
2. 14 hrs light (14 – L), and 10 hrs darkness (10 – D) at 23°C in a 'Growth chamber' (± 0.5°C).
3. 14 hrs light (14 – L), and 10 hrs darkness (10 – D) at 17°C in a 'Growth chamber' (± 0.5°C)
4. 10 hrs light (10 – L), and 14 hrs darkness (14 – D) at 23°C in a 'Growth chamber' (at ±1°C).
5. 10 hrs light (10 – L), and 14 hrs darkness (14 – D), at 17°C in a 'Growth chamber' (± 0.5°C)
6. Continuous darkness (24 – D), at 20-23°C room temperature.
7. Continuous darkness (24 – D), at 17°C, in a constant temperature room.

In each treatment large numbers of females were placed individually into jars with males after having been weighed. In *A. officinalis*, the presence of males was found not to be essential for successfull breeding largely due to the ability of females to store sperm (Warburg & Cohen 1992a,b). In *S. tiberianum*, we assumed that all collected females must have mated previously, as adult males are rare in natural populations (details in Warburg & Cohen 1991). The animals were examined weekly. Those kept in the dark were examined under red light in a dark room.

Dissections of 5-10 females were done periodically under a dissecting microscope. Both ovary and oöcyte dimensions were measured microscopically, and oöcyte numbers were counted. Some of the ovaries were then placed in suitable fixatives used for SEM examination. Some of the remaining females were kept until the marsupium was formed, and then dissected in order to establish the marsupial egg number. Other females were kept until they released their mancas.

3 RESULTS

3.1 *The breeding pattern under natural conditions*

A. officinalis breeds during fall, but can be found throughout the year. *S. tiberianum* breeds during spring, and can be found only during winter and spring. The ovary of *S.tiberianum* increased in both length and width. In *A. officinalis*, oögenesis lasted for several months, whereas in *S. tiberianum*, it lasted only for 1-2 months. Oöcyte diameter increased during oögenesis in both species. The ovaries of *A. officinalis* clearly showed large and small oöcytes (Pl.1). Its large, mature oöcytes moved into the marsupium during the breeding season. As a result, the partially empty ovarian sleeve showed some remnants which were left over from some resorbed oöcytes (Pl.1). In *S. tiberianum*, the ovary contained one row of homogeneously-sized oöcytes (Pl. 2).

Following the parturial moult when the marsupium is formed, the dimensions of the ovary (in both species) decreased and it became shorter and narrower due to the eggs moving into the marsupium. At this stage the empty ovary can be seen in both species, and in *A. officinalis* the smaller oöcytes become very clear. During this phase there was an apparent loss in its oöcyte number (Pl.1). In *A. officinalis*, the marsupium was formed in September, whereas the mancas emerged from the marsupium during September-October. In *S. tiberianum*, the marsupium was formed in March and mancas were released during March-April.

3.2 *The effect of photophase and temperature on the breeding pattern.*

In *A. officinalis* under experimental light and temperature regimes the ovaries were generally shorter and narrower than normal ovaries (Figs 1 and 2). In total darkness ovarian shape was relatively normal. Under experimental conditions other than darkness oöcyte diameter was generally smaller than that of normal oöcytes (Fig. 3). Low temperature affected oögenesis by prolonging it and causing increased oösorption, whereas high temperature shortened the duration between each phase. In all treatments the number of oöcytes was smaller than under natural conditions (Fig. 4) because of oösorption (Pl. 3). The rate of oösorption varied: It was higher at the lower temperatures (17°C), consequently oögenesis was never completed under these conditions. Marsupia were formed only at the high temperature (23°C) in light (Tables 1 and 3). Furthermore, marsupia are formed sooner than under natural conditions (Fig. 5). Moreover, the number of marsupial eggs was considerably smaller than under natural conditions, consequently fewer mancas were released, and their release occured earlier than it normally would have (Fig. 6).

In *S. tiberianum*, although oögenesis was considerably shortened, the differences in length and width of the ovary, and in diameter and number of oöcytes, between the treated and field animals were less striking than the differences found in *A. of-*

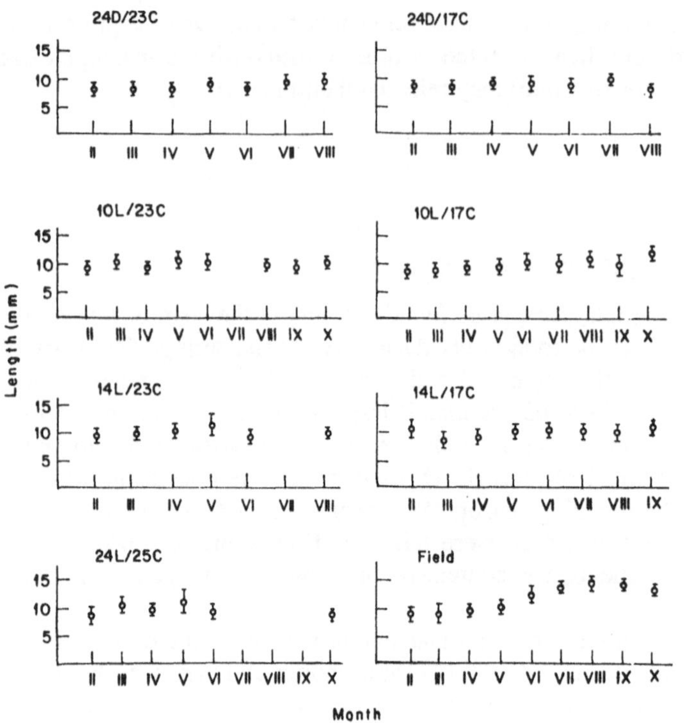

Figure 1. Average ovary length of *Armadillo officinalis.*

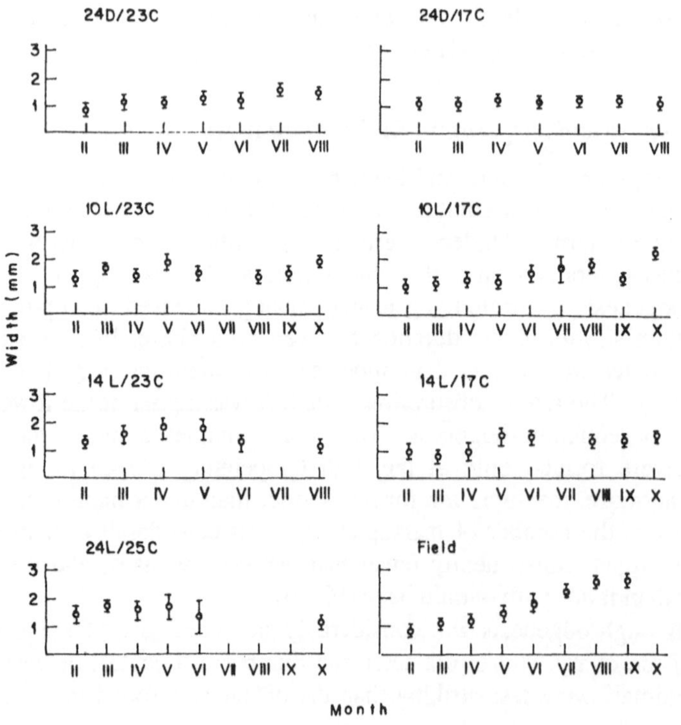

Figure 2. Average ovary width of *Armadillo officinalis.*

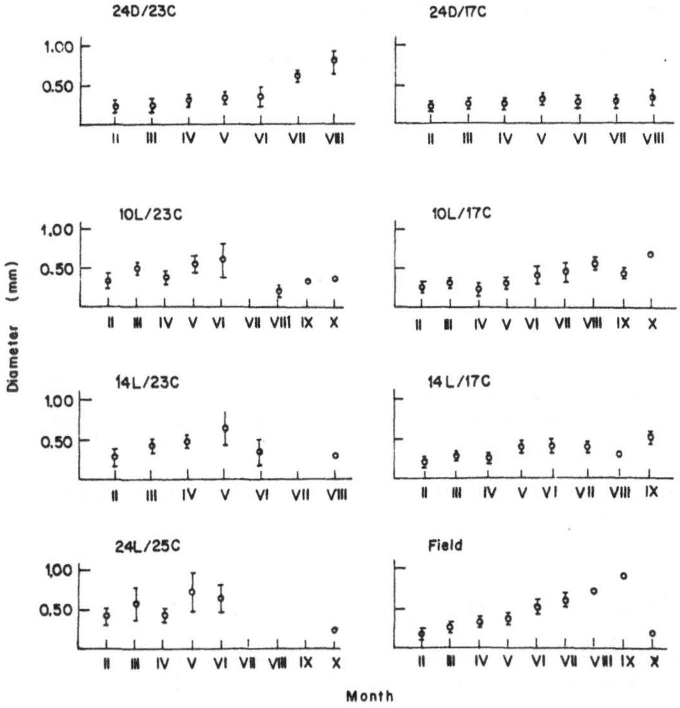

Figure 3. Average oöcyte diameter of *Armadillo officinalis*.

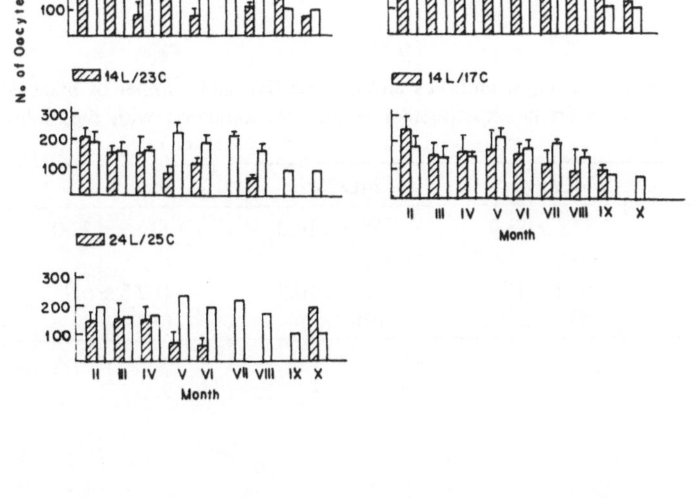

Figure 4. Average oöcyte number of *Armadillo officinalis*.

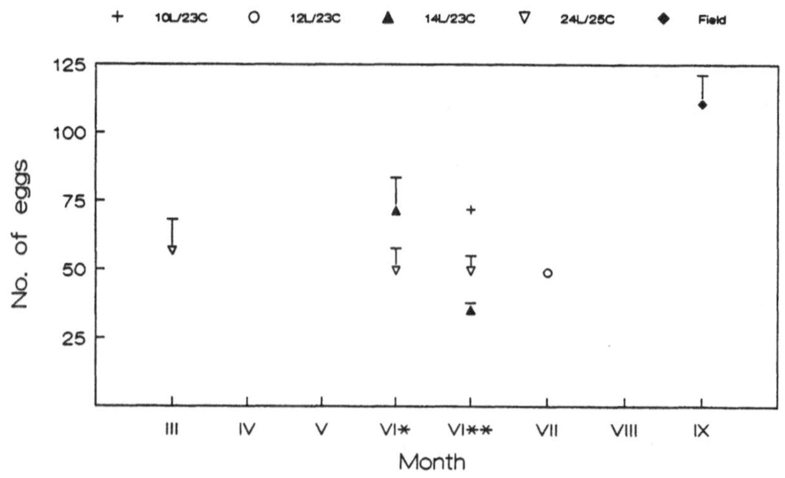

Figure 5. Marsupium formation and number of eggs in *A.officinalis* under different photophase and temperature conditions.

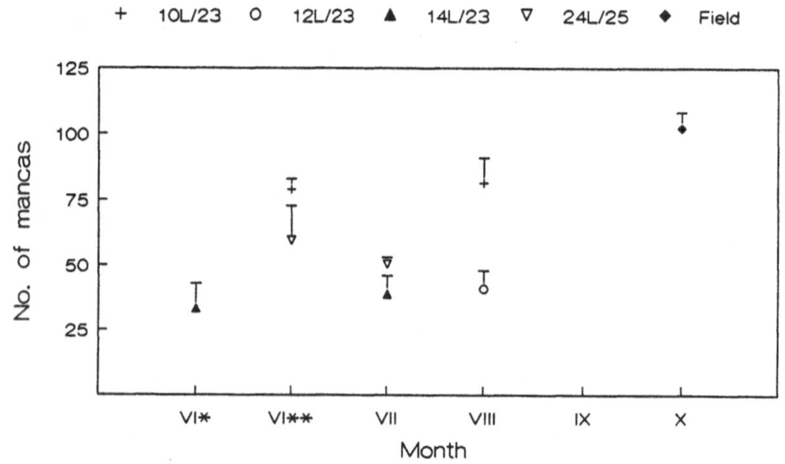

Figure 6. Manca release and number of mancas in *A. officinalis* under different photophase and temperature conditions.

Table 1. Average number of marsupial eggs, embryos and mancas (E); and number of mancas released (M) in *Armadillo* under different experimental conditions compared with the field. (n) Number of females dissected.

Treatment	10L/23C	14L/23C	24L/25C	Field
E	68.7 ± 5.7	53.5 ± 13.0	56.6 ± 14.0	112.3 ± 27.0
	(4)	(10)	(9)	(7)
M	80.0 ± 7.3	36.2 ± 8.7	55.5 ± 10.9	102.3 ± 6.0
	(4)	(4)	(9)	(5)

ficinalis (Figs 7-10), where there was a considerable loss in oöcytes (Table 3). Marsupia were formed in all treatments, although 1-3 months sooner than normally, and at a heavy price in marsupial egg loss (Tables 2 and 3). As a result considerably fewer mancas were released, and those sooner than normally (Table 2, Fig. 11). Although the rate of oösorption was lower in *S. tiberianum* compared with *A. officinalis* (Pl. 2c), the numbers of both eggs and mancas also dropped in the former species (Table 2).

4 DISCUSSION

The effects of experimentally applied photophase and temperature on the reproduc-

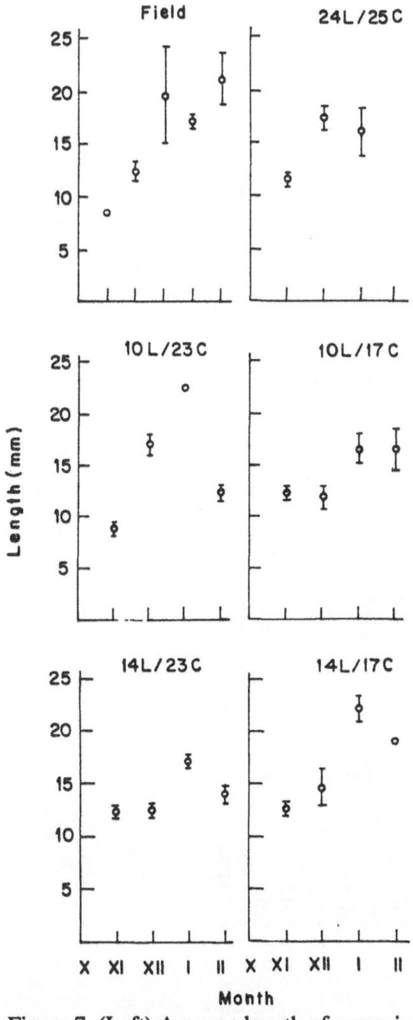

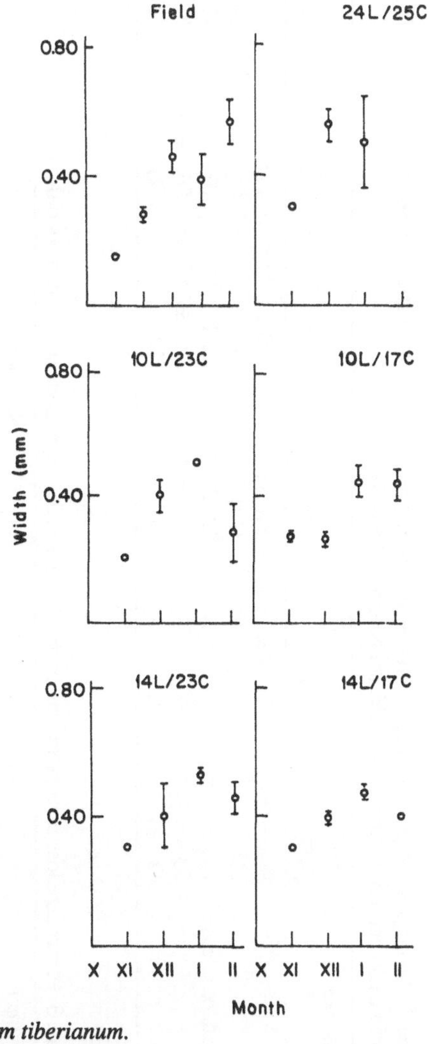

Figure 7. (Left) Average length of ovary in *Schizidium tiberianum*.
Figure 8. (Right) Average width of ovary in *Schizidium tiberianum*.

Table 2. Average number of marsupial eggs, embryos and mancas (E); and mancas released (M) in *Schizidium* during the breeding season under different experimental conditions and in the field. (n) Number of females dissected.

Treatment	17° 24D	10L	14L	23° 24D	10L	14L	25° 24L	Field
E	96.0 ± 26.0	114.7 ± 12.0	118.8 ± 25.7	100.2 ± 20.1	115.5 ± 30.5	119.5 ± 35.9	91.0 ± 36.5	154.8 ± 20.6
	(3)	(4)	(10)	(6)	(2)	(9)	(14)	(20)
Month of release	III, IV	IV	III, IV	III, IV	III	II, III, IV	II, III	IV
M	58.0 ± 16.0	40	30.3 ± 8.7	52.4 ± 24.0	56.5 ± 10.5	83.0 ± 25.0	38.4 ± 14.0	125.3 ± 9.3
	(2)	(1)	(3)	(10)	(2)	(8)	(7)	(10)

Table 3. Percentage loss in oöcytes and marsupial eggs under experimental and field conditions.

Treatment		Armadillo Oöcytes	Eggs	Schizidium Oöcytes	Eggs
Field		32.0	8.9	34.9	19.2
23°	24L	42.8	1.9	53.7	63.6
	14L	58.0	32.3	32.8	29.8
	10L	42.2	16.3	34.0	54.8
	24D	*	*	47.5	56.4
17°	14L	*	*	42.9	72.3
	10L	*	*	58.4	56.1
	24D	*	*	9.7	66.0

*Oögenesis stopped.

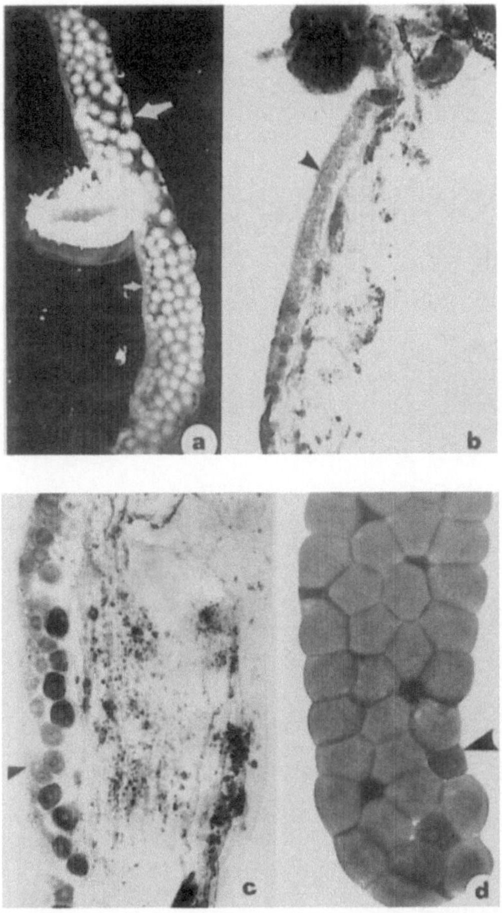

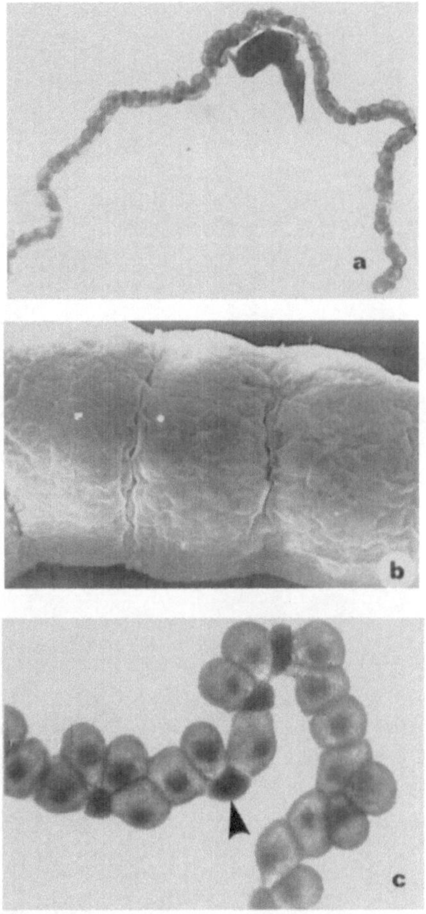

Plate 1. (Left) a) Ovary of *Armadillo officinalis*, showing large and small (arrow) oöcytes (× 9). b) Empty ovary showing a row of small oöcytes (arrow) (× 10). c) Same as in 'b' enlarged. Note two generations of the smaller oöcytes (arrow) (× 40). d) Normal oösorption (arrows) under field conditions (× 14).

Plate 2. (Right) a) Ovary of *Schizidium tiberianum*. Note the single row of homogeneously-sized oöcytes (× 12). b) Scanning electron micrograph of an ovary portion (× 200). c) Oösorption (arrows) under natural conditions (× 14).

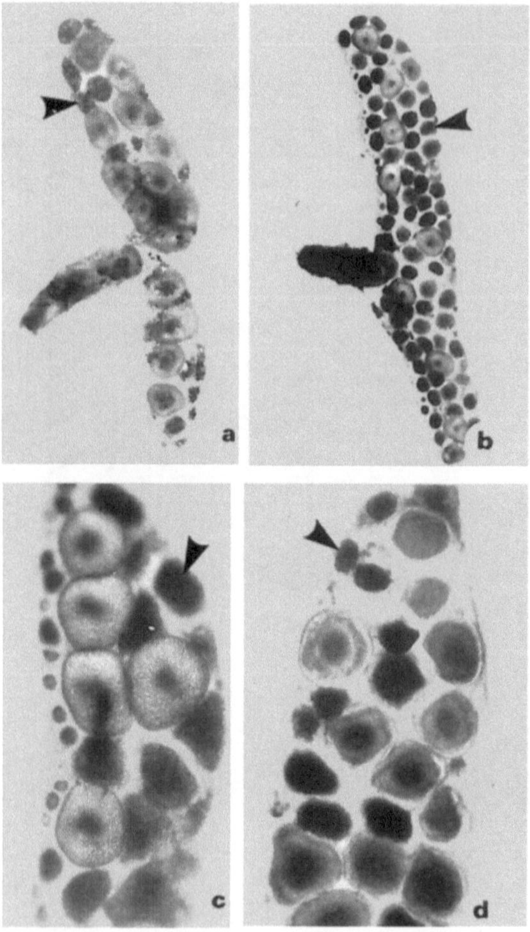

Plate 3. Oösorption (arrows) in *Armadillo officinalis*, under different temperature and photophase conditions. a) Under complete darkness (× 12). b) Under continuous light (× 9). c) Under 10 hours photophase at 23°C (× 40). d) Under 14 hours photophase at 17°C (× 50).

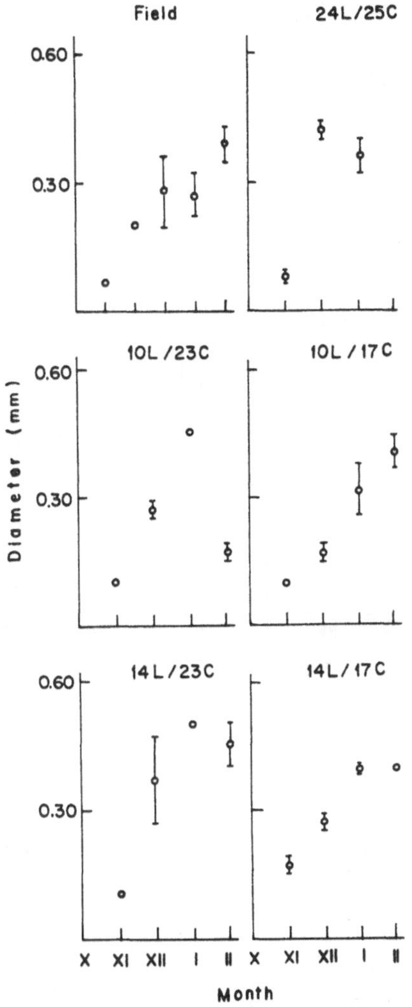

Figure 9. Average oöcyte diameter in *Schizidium tiberianum*.

tion of the two isopod species studied here indicate three points: First, these conditions produce an effect on the duration of oögenesis and formation of marsupium. Secondly, they produce an effect on the the number of oöcytes and ova being produced, and lastly, they produce an effect on the date and extent of manca release (Figs 12 and 13). Both the increased photophase and the higher temperature enhanced oöcyte resorption. Consequently, a smaller number of oöcytes matured resulting in a smaller number of marsupial ova. Oösorption was considerably more pronounced in *A. officinalis* than in *S. tiberianum* (Pl. 3).

At the lower temperature (17°C), development of the oöcytes slowed but it was not completely arrested. Marsupium formation did not occur at all photophases. Oösorption in both species resulted in fewer oöcytes, fewer ova, and fewer mancas if the latter two stages were formed.

Arrested oöcyte growth and their subsequent degeneration were previously described in *Armadillidium vulgare* when the Y-organ was extirpated (Suzuki 1986).

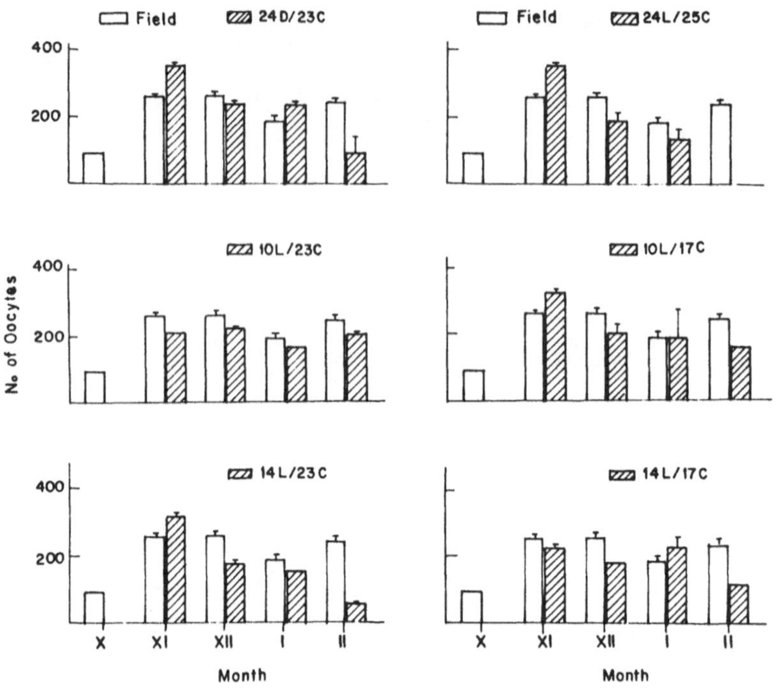

Figure 10. Average oöcyte number in *Schizidium tiberianum.*

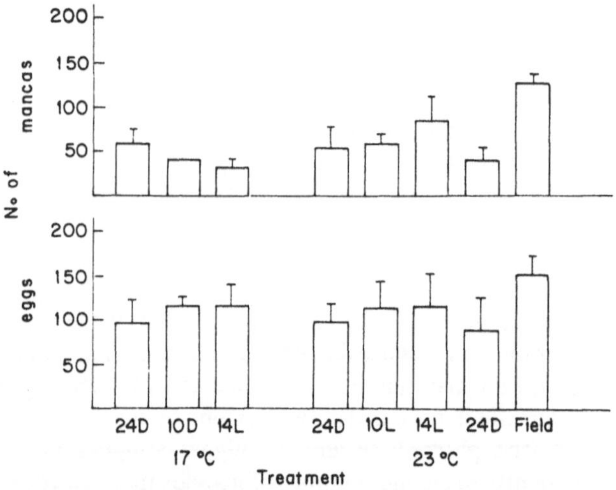

Figure 11. Number of marsupial eggs and mancas released at different photophase and temperature conditions in *Schizidium tiberianum.*

This phenomenon of oösorption is well documented in insects (Bell & Bohm 1975). Both complete darkness as well as shortened daylength are among the main causes of oösorption. In isopods it was first described in *P. ficulneus* (Hornung & Warburg 1994).

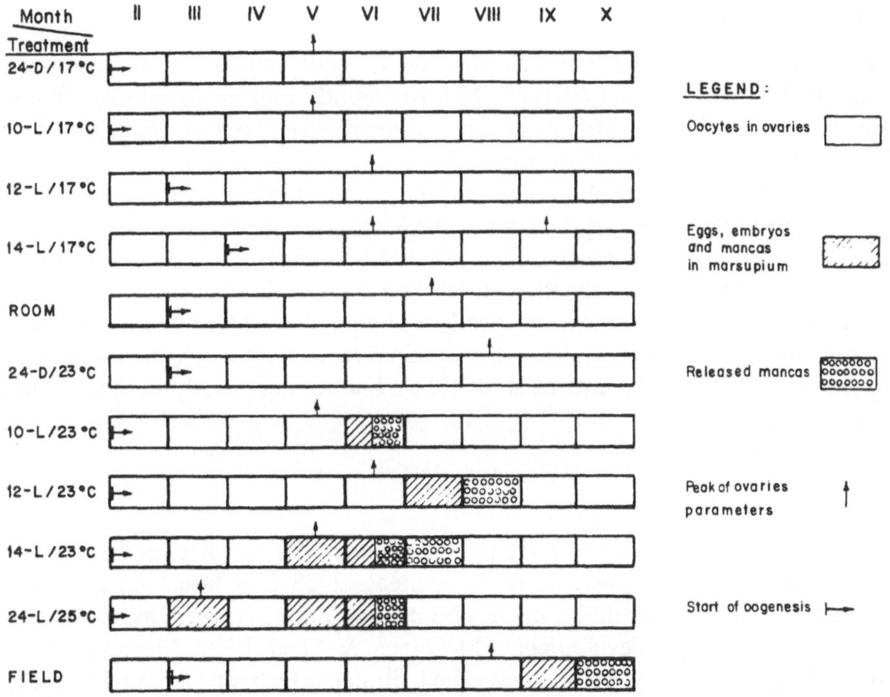

Figure 12. Reproductive patterns in *Armadillo officinalis*.

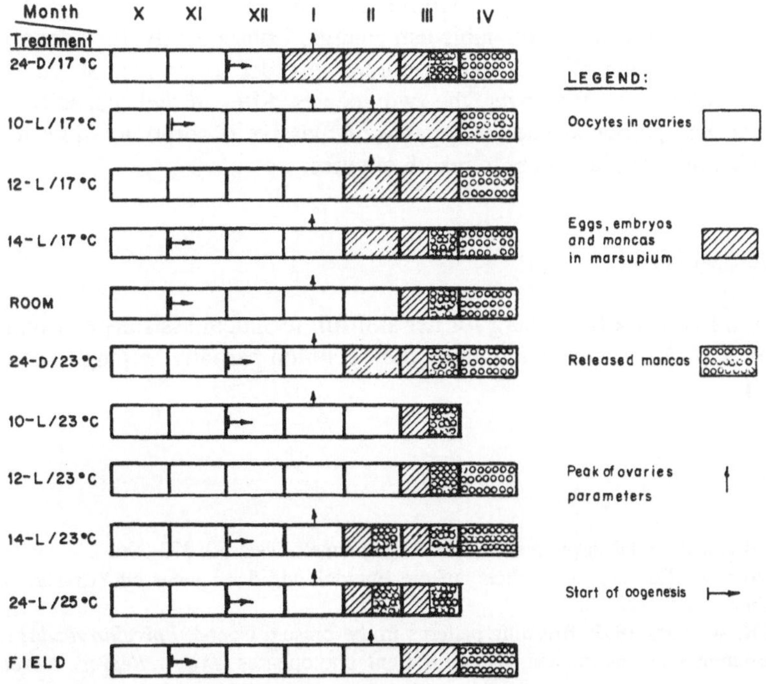

Figure 13. Same as in Figure 12 in *Schizidium tiberianum*.

4.1 *The effect of light*

Previous studies on isopods have shown that long daylength (16 hrs), initiated breeding in *Porcellio scaber* (Wieser 1963). Apparently long daylength accelerated ovarian maturation in the African species, *Eluma purpurascens* (Juchault et al. 1980). Similarly, *A. vulgare* was stimulated to breed when under constant light (Mocquard et al. 1980). Both *A. vulgare* and *P. scaber* stopped breeding under continuous darkness (Beyer 1965). Even when daylength was slightly shortened, breeding stopped in *A. vulgare* (Mocquard et al. 1980). Increased daylength caused a longer reproductive period (Juchault et al. 1981). However, in *Porcellionides pruinosus*, initiation of reproduction was independent of photoperiod (Juchault et al. 1985). It seems therefore that photoperiod is the principal factor regulating the onset of reproduction in *A. vulgare* (Mocquard et al. 1989), but not necessarily in other isopod species.

4.2 *The effect of temperature*

In *Porcellio dilatatus* the reproductive period became shorter when temperature increased. Thus at 15°C it lasted 60 days, whereas at 25°C it lasted only 25 days (Mocquard et al. 1976). Similarly, in *Oniscus asellus* and *A. vulgare*, increased temperature accelerated egg development (McQueen & Steel 1980; Mocquard et al. 1980). In the present study we have observed similar effects of increased temperature. Oögenesis became shorter thereby shortening the breeding period. On the other hand we described additional effects of increased temperature in reducing the numbers of eggs and mancas. These two effects could indicate that isopods are potentially capable of adapting to changing environment that may occur during global warming. Thus, their reproductive time-table can change. Consequently, they are apparently capable of recruiting all avaialble resources in order to produce at least a limited number of offsprings. Moreover, the two species differ in their capacity to adapt: whereas the iteroparous species is capable of massive oösorption, while the semelparous one is limited in its ability to resorb oöcytes.

ACKNOWLEDGEMENTS

We are indebted to Ms. Mira Rosenberg for her skillfull technical assistance in photography. One of us (DW), recieved a grant by the Technion partially helping to conduct this research.

REFERENCES

Bell, W.J. & M.K. Bohm 1975. Oösorption in insects. *Biological Reviews*, 50: 373-396.

Beyer, R. 1965. Uber den Einfluss des Lichtes auf die Entwicklung der Landasseln (Crustacea: Isopoda). *Pedobiologia*, 5: 122-130.

Hornung, E. & M.R. Warburg 1993. Breeding patterns in the oniscid isopod, *Porcellio ficulneus* Verh. under different temperature and under different photophases. *Invert. Reprod. Devel.* 23:151-158.

Hornung, E. & M.R. Warburg 1994. Oösorption and oöcyte loss in a terrestrial isopod under stressful conditions. *Tissue & Cell* 26:277-284.

Juchault, P., Mocquard, J.P. & S. Kouigan 1985. Etude experimentale de l'influence des facteurs externes (temperature et photoperiode) sur le cycle de reproduction du crustace oniscoide *Porcellionides pruinosus* (Brandt) provenant de populations africaine (Togo) et europeenne (France). *Crustaceana*, 48: 307-315.

Juchault, P., Mocquard, J.P. & J.J. Legrand 1981. Determinisme de la reproduction saisonniere des femelles d'*Armadillidium vulgare* Latr.(Crustace, Isopode, Oniscoide). III. Suppression ou prolongation de la periode de repos sexuel saisonnier obtenue par application de programmes photoperiodiques. *Ann. Sci. Nat., Zool., Paris,* 13 ser 3: 141-145.

Juchault, P., Mocquard, J.P., Bugrier, N. & G. Besse 1980. Croissance et cycle reproducteur du crustace isopode oniscoide *Eluma purpurascens* Budde Lund. Etude dans la nature et au laboratoire, sous differentes conditions de temperature et de photoperiode, d'une population du centre-ouest de la France. *Vie Milieu*, 30: 149-156.

McQueen, D.J. & C,G,H. Steel (1980). The role of photoperiod in the inititation of reproduction in the terrestrial isopod *Oniscus asellus* Linnaeus. *Can. J. Zool.* 58: 235-240.

Mocquard, J.P., Juchault, P. & C. Souty-Grosset 1989. The role of environmental factors (temperature and photoperiod) in the reproduction of the terrestrial isopod *Armadillidium vulgare* (Latreille,1804). *Monit. Zool. Ital. (N.S.) Monogr.* 4 : 455-475.

Mocquard, J.P., Pavese, A. & P. Juchault 1980. Determinisme de la reproduction saisonniere des femelles d'*Armadillidium vulgare* Latr.(Crustace, Isopode, Oniscoide). I. Action de la temperature et de la photoperiode. *Ann. Sci. Nat. Zool. Paris*, 14 ser. 2: 91-97.

Mocquard, J.P., Besse, G., Juchault, P., Legrand, J.J., Maissiat, J.,Martin, G. & J.L. Picaud 1976. Duree de la periode de reproduction chez les femelles de l'Oniscoide *Porcellio dilatatus* Brandt suivant les conditions d'elevage: temperature, photoperiode et groupement. *Vie Milieu* 26: 51-76.

Souty-Grosset, C., Chentoufi, A., Mocquard, J.P. & P. Juchault 1988. Seasonal reproduction in the terrestrial isopod *Armadillidium vulgare* (Latreille):Geographical variability and genetic control of the response to photoperiod and temperature. *Inver. Reprod. Devel.* 14: 131-151.

Suzuki, S. 1986. Effect of Y-organ ablation on oöcyte growth in the terrestrial isopod, *Armadillidium vulgare. Biol. Bull.* 170: 350-355.

Warburg, M.R. 1987. Isopods and their terrestrial environment. *Adv. Ecol. Res.* 17: 187-242.

Warburg, M.R. 1991. Reproductive patterns in oniscid isopods. pp 131-7 in 'Biology of Terrestrial Isopods', 3rd Int.Symp. Poitiers, eds. P.Juchault & J.P. Mocquard, Univ. Poitiers Press.

Warburg, M.R. 1994. Review of recent studies on reproduction in terrestrial isopods. *Invert. Reprod. Devel.* 24:45-62.

Warburg, M.R. & N. Cohen 1991. Reproductive pattern allocation and potential in a semelparous isopod from the Mediterranean region of Israel. *J.Crust.Biol.* 11: 368-374.

Warburg, M.R. & N. Cohen 1992a. Reproductive pattern, allocation and potential in an iteroparous isopod from a xeric habitat in the Mediterranean region. *J.Arid Environ.* 22:161-171.

Warburg, M.R. & N. Cohen 1992b. Population dynamics, growth and longevity in *Armadillo officinalis* (Isopoda;Oniscidea) inhabiting the Mediterranean region of northern Israel. *Pedobiologia* 36:262-273.

Warburg, M.R., Linsenmair, K.E. & K. Bercovitz 1984. The effect of climate on the distribution and abundance of isopods. *Symp. Zool. Soc. Lond.* 53: 339-367.

Wieser, W. 1963 Die Bedeutung der Tageslange fur das Einsatzen der Fortpflanzungsperiode bei *Porcellio scaber* Latr. (Isopoda). *Z. Naturf.* 18: 1090-1092.

The effect of leaf litter, microorganisms and Collembola on the food allocation of *Oniscus asellus* (Isopoda)

M. Uesbeck & W. Topp
Department of Zoology, University of Cologne, Köln, Germany

ABSTRACT

Oniscus asellus was fed alternatively with birch- or oak litter. The feeding performance showed that relative growth rates (= RGR) and efficiencies in converting ingested food (= ECI) and digested food (= ECD) into body mass as well were more affected by the food items than by the microflora. In contrast, the increase of approximate digestibility (= AD) by microorganisms, probably via their cellulase activity, was striking. Microflora, at an elevated density, increased AD-values from 45% to about 75% on birch, and oak litter as well. Additionally the microflora stimulated the increase of the consumption rates (RCR). On oak litter a 6-fold increase of RCR was observed. Increased food consumption is interpreted as a response to suboptimal nutrient levels in leaf litter to prevent large reductions in growth. When feeding *O. asellus* on birch litter there is enhanced cellulase activity but not dehydrogenase activity. On oak, litter enzyme activity was not affected by this isopod. These data indicate that *Oniscus asellus* benefits by stimulating the microbial activity. Also, grazers such as the Collembola *Folsomia candida* improve the nutritional values of the litter consumed by *Oniscus asellus*. This was demonstrated for birch litter where *F. candida* enhanced both dehydrogenase and cellulase activity. The analyses highlight the importance of interactions among microorganisms, microarthropods, and litter quality.

1 INTRODUCTION

Isopods are considered as generalist feeders that are detritovorous or omnivorous animals feeding on live and dead leaves, fungi, live or dead animals, and even on their own fecal pellets (Warburg 1987). Although mostly omnivorous, individual species exhibit marked preferences for particular types of leaf litter (Rushton et al. 1983, Dudgeon et al. 1990). *Oniscus asellus* usually prefers lime, ash, and alder leaves rather than leaves of oak and beech (Dunger 1962). Moreover, Dallinger et al. (1977) found that assimilation increased with the concentration of copper ions in the food. Similarly, the desert woodlouse *Hemilepistus reaumuri* prefers *Hammada* and *Artemisia*. But in order to survive this isopod should include soil particles in its nourishment (Shachak et al. 1976).

Several studies on aquatic arthropods have demonstrated a preference for litter that supports a rich culture of microorganisms (Anderson et al. 1979; Cummins et al. 1979). This preference for microbially colonized substrates has been interpreted either as a selection for microbial tissue, which is rich in minerals, micronutrients, and easily assimilated protein, or as a selection to acquire hydrolytic enzymes of microbial origin. The ingestion of microbial enzymes plays a dominant role in some termites and other wood living insects (Kukor et al. 1983). The same may be valid for woodlice. It has been shown that woodlice are able to digest hemicellulose but not cellulose (Reyes et al. 1976), and that ingested microbial enzymes (Hassall et al. 1975) and bacteria are present in the gut fluid. The dominating microbes in the intestine and on the feces of *O. asellus* in contrast to the litter are coryneform bacteria. These bacteria are known as suppliers of extracellular enzymes and essential amino acids (Ullrich et al. 1991). The amphipod, *Gammarus fossarum*, is able to digest cellulose and there is evidence that cellulolytic enzymes present in the gut fluids originate from the food (Bärlocher 1982). Also the studies carried out with the common woodlouse *Tracheoniscus rathkei* by Kukor et al. (1986) clearly demonstrate the potential contribution of acquired enzymes to the digestion of cellulose. Using microcosms which contained decomposing *Pinus nigra* litter Teuben et al. (1990) demonstrated that the fungivorous collembolan *Tomocerus minor* and the detritovorous isopod *Philoscia muscorum* are each able to stimulate microbial activity including cellulase activity. In this paper, the effects of the collembolan, *Folsomia candida*, and the isopod *Oniscus asellus*, on microbial activity and especially on cellulase activity were tested. In order to find out whether microbial activity was enhanced by a mutual effect between both arthropods, they were treated separately and in common. At the same time as the experiments for testing interactions between microbial activity and arthropods, further experiments were carried out to obtain consumption rates of *Oniscus asellus*. The experimental set-up was chosen to address the following questions:

1. Is the feeding performance of *Oniscus asellus* exclusively dependent on the litter quality?
2. Is there any effect of the microorganisms on the feeding performance of *O. asellus*?
3. Is the growth of microorganisms stimulated by the activity of *O. asellus* or the grazer *Folsomia candida*?
4. Are the woodlice able to benefit from the activity of the Collembola in respect to the feeding performance?

2 MATERIALS AND METHODS

Leaf litter and animals were sampled in October in a broadleaved mixed forest situated near the river Rhine (51°46'N 6°24'E). The leaf litter used for the experiments was analyzed for parameters that may influence microbial and animal activity.

Water content: Leaf litter was placed over a saturated atmosphere over a time period of 24 h then dried to constant weight in an oven at 60°C for 24 h. Percentage of leaf water of control leaves was determined from the difference in the two weights.

Toughness: The toughness of the leaf litter was estimated with a penetrometer similar to the one used by Feeny (1970). This device measures the weight required to drive a small piston through a leaf.

Nitrogen-content: Nitrogen measurements were Kjeldahl-N.

Tannin-concentration: Condensed tannins were estimated as proanthocyanidins. Hydrolyzable tannins were measured using HCl – nBuOH solutions (Bate-Smith 1977, Stafford et al. 1986). Data are reported colorimetrically (E_{550}) derived from standard curves.

Microbial activity was measured in terms of dehydrogenase and cellulase activity. Dehydrogenase activity was measured according to Thalmann (1968) whereas cellulase activity was measured by a method developed by Nelson (1944) and modified by Kühle et al. (1990). The substrates tested were fragmented birch and oak leaf litter separated in samples of 500 mg. They were defaunated and partly sterilized using a heating-thawing-remoistening-freezing treatment, similar to Huhta et al. (1990). Thereafter leaf litter was remoistened again and microbial populations were reinoculated from a litter suspension obtained from litter that was collected at the same time and the same sites as the substrates.

Reinoculation and subsequent development occurred in cloths with 5 µm mesh size over a time period of three weeks. After this time interval samples were placed into plastic boxes with a bottom covered by a texture of 1 mm Ø to allow the faeces to fall into a second box covered by a texture of 150 µm Ø. From this box the faeces were sampled. The procedure was chosen to minimize coprophagy. Nevertheless, single faeces remained on the texture and the leaf litter. Both boxes were incubated into a third one that was tightly closed and that contained a water bath on its bottom. So woodlice lived and fed at saturated humidity. Each trial contained 20 parallels. The experiments were carried out at 15°C and short day conditions (= LD 8/16).

For calculating feeding performances the defaunated and partly sterilized leaf litter and the leaf litter on which the microflora developed over three weeks were used. The 'sterilisation' procedure used was developed to defaunate and partly sterilize soil samples. The method also seeks to study the effect of microorganisms on leaf litter. Immediately after the 'sterilization' process of leaf litter no microbial activity according to the methods mentioned above could be measured. Estimates of growth and feeding for *Oniscus asellus* are based on dry weight measurements. Values were obtained after a feeding period of 3 weeks using individuals with an initial wet weight of 20-60 mg. Initial dry weight measurements for the individuals used in the experiments were calculated by a regression line dry weight vs. fresh weight ($y = 0.28 \times - 0.001$, $r = 0.994$, $p < 0.001$). From our data we calculated the following parameters of growth and feeding efficiency (Waldbauer 1968):

R.G.R.: Relative Growth Rate
= (mg gained/mg mean biomass) × day

R.C.R.: Relative Consumption Rate
= (mg eaten/mg mean biomass) × day

A.D.: Approximate Digestibility
(assimilation efficiency)
= [(mg eaten-mg faeces)/mg eaten] × 100

E.C.D.: Efficiency of Conversion of Digested food
(net growth efficiency)
= [mg biomass gained/(mg eaten-mg faeces] × 100

E.C.I.: Efficiency of Conversion of Ingested food
(gross growth efficiency)
= (mg biomass gained/mg eaten) × 100

3 RESULTS

3.1 *Characterisation of leaf litter*

Some of the key litter properties to which consumers have evolved different morphological, behavioural and/or physiological traits are summarized in Table 1. One single factor or the combined effect of some of the properties measured may influence growth of microorganisms. Also they directly affect the reaction pattern of arthropods or they may influence indirectly the traits of the consumers by the meaning of microorganisms. The factors in table 1 which characterize the quality of leaf litter from *Betula pendula* on the one hand and the leaf litter from *Quercus robur* on the other all were significantly different (t-test, $p < 0.001$).

3.2 *Leaf litter suitability and the effect of microbial activity on the feeding performance*

Growth rates are derived from feeding rates, reflecting behavioural responses of isopods to food, and from feeding efficiencies, which represent metabolic responses to ingested food (RGR = RCR × ECI). The faster growth of *Betula pendula* isopods on incubated instead of sterilized leaf litter seems to have been due partly to faster feeding and partly to the greater efficiency at digested food (Table 2). The somewhat higher values of ECI (ECI = AD × ECD) when feeding on incubated leaf litter may be due mainly to the substantially higher digestibility (AD) in comparison to the sterilized leaf litter (t-test, $p < 0.001$).

Microbial activity on *Quercus robur* leaves may have the same influence on the feeding parameters as on *Betula pendula* leaves. The substantially higher digestibil-

Table 1. Factors characterizing the quality of leaf litter (× ±SD).

	Betula pendula			*Quercus robur*		
	n	×	±SD	n	×	±SD
Water content (% dry wt at saturated humidity)	10	69.40	2.60	10	60.50	2.20
Nitrogen content (% dry wt)	7	2.44	0.08	5	0.84	0.08
Toughness (g/cm^2)	6	363.90	67.60	6	743.80	35.50
Soluble tannins (100 × E/mg × ml)	8	0.04	0.01	6	0.44	0.12
Condensed tannins (E/mg × ml)	7	0.20	0.04	7	0.32	0.07

Table 2. Rates and efficiencies of food consumption and utilisation for *Oniscus asellus* in relation to leaf litter and microbial activity. Microbial activity of the incubated leaf litter corresponds to the data in Tables 5 and 6 (series I) for leaf litter plus two *O. asellus* (n = 20 for each trial: × ±SE).

		RCR (mg/g × d)	RGR (mg/g × d)	AD (%)	ECI (%)	ECD (%)
B. pendula	n	15	15	15	15	15
Incubated	×	126.5^{a}	6.6^{a}	72.4^{a}	6.4^{a}	7.1^{a}
Leaf litter	SE	15.1	1.7	1.6	1.8	1.8
B. pendula	n	11	11	11	11	11
Sterilized	×	92.8^{a}	3.0^{ac}	46.3^{b}	4.0^{ac}	5.0^{a}
Leaf litter	SE	11.2	0.6	5.7	0.9	1.2
Q. robur	n	10	10	10	10	10
Incubated	×	201.0^{b}	0.6^{bc}	78.7^{a}	0.5^{b}	0.7^{b}
Leaf litter	SE	24.8	1.5	1.6	0.8	0.9
Q. robur	n	9	9	9	9	9
Sterilized	×	33.9^{c}	0.3^{b}	45.5^{b}	0.8^{bc}	0.9^{b}
Leaf litter	SE	6.8	0.4	7.8	1.7	1.8

[a]Means in columns not followed by the same letter are significantly different.

ity (AD) may cause the tendency of an increasing RGR when feeding on the incubated leaf litter.

A comparison of the leaf litter from the two trees showed for the incubated as well as the sterilized litter a substantially slower growth rate (t-test, $p < 0.01$) when isopods fed on *Quercus robur* leaves (Table 2). These differences were caused by the extremely low ECI values. The relative consumption rates decreased as long as sterilized leaf litter from *Q. robur* was offered. In contrast, when incubated leaf litter from *Q. robur* was offered, consumption rates of the isopods increased.

However, the low efficiency of metabolism of ingested food (ECI) was not completely offset by the high RCR to result in a growth rate comparable to the rates obtained from *B. pendula* leaf litter. Such reaction pattern was expected because in isopods as in phytophagous insects an increased food consumption may be a response to suboptimal nutrient levels in foliage and because feeding efficiencies tend to be inversely correlated with feeding rates (Tabashnik et al. 1987; Topp et al. 1989). AD values for the isopods feeding either on incubated or on sterilized litter were almost identical for both food items. Microbial activity explained as much as 43% of the variance measured in assimilation efficiency. An analysis of variance comparing the four experimental series indicated that, although consumption rate and approximate digestability were mostly affected by microbial activity, leaf litter had a greater impact on growth rate and on food conversion efficiencies. Furthermore, there were significant leaf litter/microbial activity interactions on RCR. However, no significant leaf litter/microbial activity interactions for RGR were obtained, suggesting that isopods exhibiting low growth rates because of the low nutritional value of the food did so independently of the microbial activity (Table 3).

Table 3. Analysis of variance on feeding parameters of *Oniscus asellus* when fed on leaf litter from *Betula pendula* and *Quercus robur*.

Source	SS	df	MS	F	SS	df	MS	F
	RCR				RGR			
Leaf litter	0.00	1	0.04	1.2	0.00	1	0.00	11.3[1]
Microbial activity	0.10	1	0.10	33.2[3]	0.00	1	0.00	2.9
Interaction (leaf litter × microbial activity)	0.54	1	0.54	18.2[3]	0.00	1	0.00	1.6
Error	0.12	41	0.00		0.00	41	0.00	
	ECI				ECD			
Leaf litter	238.02	1	238.02	10.9[2]	315.67	1	315.67	11.2[2]
Microbial activity	16.70	1	26.70	0.8	13.21	1	13.21	0.5
Interaction (leaf litter × microbial activity)	17.90	1	17.90	0.8	14.48	1	14.48	0.5
Error	893.93	41	21.80	1153.16	41	28.13		
	AD							
Leaf litter	77.43	1	77.43	0.3				
Microbial activity	9741.15	1	9741.15	43.0[3]				
Interaction (leaf litter × microbial activity)	112.75	1	112.75	0.5				
Error	9275.09	41	226.22					

[1] $p < 0.05$; [2] $p < 0.01$; [3] $p < 0.001$.

3.3 *Interactions between a grazer and microorganisms*

Table 4 summarizes the experiments that were carried out to demonstrate the effect of a grazer as *Folsomia candida* on microbial activity. In most trials *F. candida* had a positive effect on enzyme activity. When 10 individuals of *F. candida* were added to the leaf litter of *B. pendula*, dehydrogenase activity increased by approximately 50% (KS-test, $p < 0.05$) in comparison to the control series whereas cellulase activity increased by nearly 100% (KS-test, $p < 0.001$). As population density *of F. candida* increased, enzyme activity decreased. On comparing cellulase activity between the control series and the series in which 20 individuals of *F. candida* were added, differences are still significant (KS-test, $p < 0.01$).

However, a population density of 40 collembolans did not show any more obvious differences in enzyme activity in comparison to the control series. The same effects – interaction between a grazer and microorganisms and the density dependence of a grazer – should also occur when leaf litter from *Quercus robur* is used instead of leaf litter from *Betula pendula*. However, the different trials showed no significant differences although carried out under identical conditions (KS-test, $p > 0.05$).

3.4 *Interactions between a grazer, microorganisms and Oniscus asellus*

The series mentioned in Table 4 and the series I of Table 5 were carried out simultaneously. As shown in Table 5 dehydrogenase activity increased. Significant differences to the control occurred only when 2 isopods and 10 collembolans where added to the leaf litter (KS-test, $p < 0.05$).

Comparing the trials in Table 4 when 10 or 20 individuals of *F. candida* were added with the trials in Table 5 (I) when in addition to the collembolans 2 individuals of *O. asellus* had the opportunity to feed on the same leaf litter no differences in dehydrogenase activity were obvious. When solely the 2 isopods fed on the leaf litter of *B. pendula* dehydrogenase activity increased slightly (KS-test, $p > 0.05$). Thus, *O. asellus* probably does not influence total microbial activity.

Table 4. Microbial activity (× ±SE) expressed as dehydrogenase activity and cellulase activity on the leaf litter from *B. pendula* and *Q. robur* respectively when measured after 6 weeks of incubation. The collembolans (*Folsomia candida*) were added after 3 weeks of incubation.

Activity	Dehydrogenase activity			Cellulase		
	n	×	±SE	n	×	±SE
Betula pendula						
Control without *F.c.*	7	201.9[a]	18.9	5	71.3[a]	8.5
10 *F. candida*	7	305.6[b]	36.4	5	138.6[c]	2.1
20 *F. candida*	10	266.1[a]	28.3	15	104.6[b]	7.2
40 *F. candida*	10	198.1[a]	23.7	15	90.2[a]	8.8
Quercus robur						
Control without *F.c.*	4	301.0[a]	58.8	5	59.7[a]	3.8
10 *F. candida*	8	382.7[a]	63.9	5	75.6[a]	5.0
20 *F. candida*	10	413.8[a]	61.9	5	83.7[a]	7.7
40 *F. candida*	10	256.3[a]	46.6	5	64.2[a]	4.7

[a]Means in columns not followed by the same letter are significantly different.

Table 5. Microbial activity (×±SE) expressed as dehydrogenase activity and cellulase activity on the leaf litter from *B. pendula* measured after 6 weeks of incubation. The isopods (*Oniscus asellus*) and the collembolans (*Folsomia candida*) were added after 3 weeks of incubation. Series I were measured in the middle of November, series II in the middle of February.

Activity	Dehydrogenase activity			Cellulase		
	n	×	±SE	n	×	±SE
I						
Control, without arthropods	7	201.9[a]	18.9	5	71.3[a]	8.5
+ 2 *O. asellus*	8	270.6[a]	38.0	5	113.2[b]	11.9
+ 2 *O. asellus* + 10 *F.candida*	9	304.2[b]	13.9	5	69.0[a]	3.5
+ 2 *O. asellus* + 20 *F. candida*	9	246.7[a]	25.0	5	61.5[a]	2.8
II						
Control, without arthropods	10	157.5[a]	29.9	5	7.2[a]	0.9
+ 2 *O. asellus*	10	122.7[a]	14.6	5	24.4[c]	1.0
+ 2 *O. asellus* + 10 *F. candida*	9	121.5[a]	21.2	5	12.4[b]	1.2
+ 2 *O. asellus* + 20 *F. candida*	8	108.6[a]	11.4	5	10.6[b]	1.5

[a]Means in columns for the series I and II respectively not followed by the same letter are significantly different.

Table 6. Microbial activity (×±SE) expressed as dehydrogenase activity and cellulase activity on the leaf litter from Quercus robur measured after 6 weeks of incubation. Further information see Table 3.

Activity	Dehydrogenase activity			Cellulase		
	n	×	±SE	n	×	±SE
I						
Control, without arthropods	4	301.0[a]	58.8	5	59.7[a]	3.8
+ 2 *O. asellus*	4	365.7[a]	128.9	5	47.3[a]	0.6
+ 2 *O. asellus* + 10 *F. candida*	9	272.0[a]	42.5	5	58.8[a]	2.1
+ 2 *O. asellus* + 20 *F. candida*	8	195.3[a]	37.9	5	61.8[a]	2.3
II						
Control, without arthropods	10	146.7[a]	18.7	5	17.4[a]	2.4
+ 2 *O. asellus*	5	108.3[a]	43.4	5	5.8[b]	1.8
+ 2 *O. asellus* + 10 *F. candida*	10	97.7[a]	10.8	5	3.9[b]	0.6
+ 2 *O. asellus* + 20 *F. candida*	10	109.2[a]	21.6	5	2.9[b]	0.2

[a]Means in columns for the series I and II respectively not followed by the same letter are significantly different.

With respect to cellulase activity the two isopods had a significant positive effect (KS-test, $p < 0.01$ in series I and $p < 0.001$ in series II). In series II (Table 5) an increase in cellulase activity was also obvious when isopods and collembolans fed in concert. The trials using leaf litter from *Quercus robur* as food items did not differ in most cases (KS-test, $p > 0.05$). An exception were the results in series II (Table 6). Here a decrease in cellulase activity occured (KS-test, $p < 0.05$ and $p < 0.01$) as soon as the arthropods were added to the leaf litter. The controls from series I and series II differed from each other. This was apparent for both food items. It is premature to interprete these differences as a seasonal effect.

4 DISCUSSION

4.1 *Feeding performance*

Food intake, utilization, and allocation are active, dynamic processes that can be altered by an organism in an adaptive manner in response to extrinsic and intrinsic factors (Scriber et al. 1981). Several species of insects have been shown to increase consumption on foods with reduced nutrient levels. In contrast, a highly digestable food may result in a reduction in RCR. Furthermore, a low RCR may increase AD because of prolonged retention of food in the gut. These trade-offs in food consumption and in food utilization efficiencies may allow a species to maintain its growth rate. Or at least, the decrease of growth rate may be lessened compared with the outcome which would have occurred if the species did not increase its consumption rate or efficiencies. Some species appear to be better compensators than others (Slansky et al. 1989; Kirsten et al. 1991). Because of these alterations it is often difficult to assess cause-effect relationships. Similar reaction patterns as known from herbivorous insects also may occur in detritivorous soil animals.

Trade-offs which may have been provoked by the reduced litter quality found in oak leaves (Table 1) give an explanation for the differences in consumption rates (RCR) obtained from leaf litter used in this study. Consumption rates on sterilized litter from oak leaves were significantly less than on sterilized litter from birch leaves ($p < 0.001$, Table 2). In contrast, when incubated litter was offered consumption rate on oak leaves gave a 6-fold increase and even exceeded the values obtained from birch leaf litter ($p < 0.01$, Table 2). Similar observations were made in earlier studies. Wieser (1965) and Dallinger et al. (1977) reported that isopods are hyperphagous when feeding on dried litter not in an advanced state of decay and explained this behaviour by the teleological advantage that organic material is more rapidly converted into faeces which provide an environment prepared for the attack of microorganisms.

Microorganisms may be crucial for the nutrition and development of macroarthropods by supporting the saprophagic species with enzymes, vitamins and amino acids. Especially the purpose of the coprophagous behaviour in terrestrial woodlice can be attributed to gain these essential components (Ullrich et al. 1991). An enzyme which can be found in the gut fluid of isopods is cellulase. There is convincing evidence that cellulase is derived from the food (Hassall et al. 1975) and that ingested bacteria, which are known as suppliers of extracellular enzymes, are present and active in the gut fluids (Reyes et al. 1976; Ullrich et al. 1991). Kukor et al. (1986) gave a first direct demonstration that assimilation efficiency (AD) in a detritivore can be enhanced by the ingestion of microbial enzymes. In their study, Kukor et al. (1986) fed *Tracheoniscus rathkei* with wheat straw amended with a commercial preparation of the cellulase complex from a fungus, *Penicillium funicolosum* and obtained an increase in AD of about 20% in comparison to wheat straw not amended with cellulase.

In our studies the assimilation efficiency of *O. asellus* increased by about 30% in comparison to sterilized leaf litter when the isopods were provided with leaf litter which was supported with microorganisms. Furthermore, from the results on cellulase activity on birch leaves (Table 5) it is supposed that isopods are able to increase the cellulolytic microbes rather than the whole spectrum of microorganisms. Increasing the abundance of cellulolytic microorganisms may only occur on appropriate substrats but not on all leaf litter items (Table 6).

Testing the total effect of microorganisms on growth rates (RGR) of *Oniscus asellus*, we found that litter quality, when characterized by higher values in water and nitrogen content but lower values in toughness, soluble and condensed tannins, will have a superior effect than microbes. Microbes affected growth rates by increasing assimilation efficiencies and probably by increasing consumption rates (recall that $RGR = RCR \times AD \times ECD$ and that $ECI = AD \times ECD$).

4.2 *Interactions*

Soil arthropods and decomposing microorganisms interact in a complex way. Factors affecting these interactions include the properties of the substrate, the successional attributes during the process of decomposition, the feeding habits of specific soil arthropods, the density of soil arthropods, and the preference for available food items.

For the experimental set-up used in this study a low density of a grazer (= 10 individuals of *Folsomia candida*) feeding on leaf litter from *B. pendula* resulted in a substantial increase of microbial activity as measured by dehydrogenase and cellulase activity. Instead, the same low density of *F. candida* feeding on *Q. robur* leaves enhanced dehydrogenase activity only slightly. These differences may be due to litter quality (Table 1) or brought about by the activities measured in the control series. Teuben (1991) stated that feeding activities by the animals increased microbial activity when control levels were low but had negative effects when control values were high.

An increase in the density of the grazer from 10 individuals of *F. candida* up to 40 individuals of *F. candida* feeding on 500 mg substrate resulted in a decrease of microbial activity. Similar reaction patterns of grazers on microbial activity were found by Teuben et al. (1990) for the fungivorous collembolan *Tomocerus minor* feeding on pine litter, by van der Drift et al. (1977) for the collembolan *Onychiurus quadriocellatus* feeding on pellets from diplopods and also by Hanlon et al. (1979) using *F. candida* and oak litter in their study. Overpopulation lessening the microbial activity was also mentioned by Ineson et al. (1982).

Unlike *F. candida* the isopod *Oniscus asellus* which mostly comminutes its food resource did not enhance dehydrogenase activity on either substrate. Similar to *F. candida* cellulase activity was enhanced on leaf litter from *B. pendula* but not on leaf litter from *Q. robur*. These results partly confirm the data obtained by *Philoscia muscorum* which were carried out in microcosm experiments (Teuben et al. 1990). Measuring fungal and bacterial standing crops on fragmented oak leaves Hanlon et al. (1980) found a decrease in fungal standing crop but an increase in bacterial standing crop when under the influence of *O. asellus*.

The common influence of a grazer (*F. candida*) and a comminutor (*O. asellus*) enhanced dehydrogenase activity almost to the same degree as low density of *F. candida* does alone (Table 3). There was no common effect of both arthropods increasing cellulase activity. The differences between the alterations of dehydrogenase – and cellulase activity under the influence of both arthropods remain unclear. The Collembola may select those cellulolytic microorganisms that are promoted by the isopods or those cellulolytic microorganisms that are enhanced by activity of isopods are more sensitive to mechanical damage than other microbes.

REFERENCES

Anderson, N.H. & J.R. Sedell 1979. Detritus processing by macroinvertebrates in stream ecosystems. *Annu. Rev. Entomol.* 24: 351-377.

Bärlocher, F. 1982. The contribution of fungal enzymes to the digestion of leaves by *Gammarus fossarum* Koch (Amphipoda). *Oecologia* 52: 1-4.

Bate-Smith, E.C. 1977. Astringent tannins of *Acer* species. *Phytochemistry* 16: 1421-1426.

Cummins, K.W. & M.J. Klug 1979. Feeding ecology of stream invertebrates. *Annu. Rev. Ecol. Syst.* 10: 147-172.

Dallinger, R. & W. Wieser 1977. The flow of copper through a terrestrial food chain. I. Copper and nutrition in isopods. *Oecologia* 30: 253-264.

Drift, J. van der & E. Jansen 1977. Grazing of springtails on hyphal mats and its influence on fungal growth and respiration. Soil organisms as components of ecosystems. *Ecol. Bull.* 25: 203-209.

Dudgeon, D., H.H.T. Ma, & P.K.S. Lam 1990. Differential palatability of leaf litter to four sympatric isopods in a Hong Kong forest. *Oecologia* 84: 398-403.

Dunger, W. 1962. Methoden zur vergleichenden Auswertung von Fütterungsversuchen in der Bodenbiologie. *Abh. u. Ber. Naturkundemus. Görlitz* 37: 143-162.

Feeny, P.P. 1970. Seasonal changes in oak leaf tannins and nutrients as a cause of spring feeding by winter moth caterpillars. *Ecology* 51: 565-581.

Hanlon, R.D.G. & J.M. Anderson 1979. The effects of Collembola grazing on microbial activity in decomposing leaf litter. *Oecologia* 38: 93-99.

Hanlon, R.D.G. & J.M. Anderson 1980. Influence of macroarthropod feeding activities on microflora in decomposing oak leaves. *Soil. Biol. Biochem.* 12: 255-261.

Hassall. M. & J.B. Jennings 1975. Adaptive features of gut structure and digestive physiology in the terrestrial isopod *Philoscia muscorum* (Scopoli) 1763. *Biol. Bull.* 149: 348-364.

Hassall, M. & P. Rushton 1985. The adaptive significance of coprophagous behaviour in the terrestrial isopod *Porcellio scaber. Pedobiologia* 28: 169-175.

Huhta, V. & H. Setälä 1990. Laboratory design to simulate complexity of forest floor for studying the role of fauna in the soil processes. *Biol. Fertil. Soils* 10: 155-162.

Ineson, P., M.A. Leonard, & J.M. Anderson 1982. Effects of collembolan grazing upon nitrogen and cation leaching from decomposing leaf litter. *Soil. Biol. Biochem.* 14: 601-605.

Kirsten, K. & W. Topp 1991. Acceptance of willow-species for the development of the winter moth, *Operophtera brumata* (Lep., Geometridae). *J. Appl. Ent.* 111: 457-468.

Kühle, K. & W. Kleinow 1990. Glucosidases in *Brachionus plicatilis* (Rotifera). *Comp. Biochem. Physiol.* 95B: 393-402.

Kukor, J.J. & M.M. Martin 1983. Acquisition of digestive enzymes by siricid woodwasps from their fungal symbiont. *Science* 220: 1161-1163.

Kukor, J.J. & M.M. Martin 1986. The effect of acquired microbial enzymes on assimilation efficiency in the common woodlouse, *Tracheoniscus rathkei. Oecologia* 69: 360-366.

Nelson, N. 1944. A photometric adaptation of the Somogyi method for the determination of glucose. *J. Biol. Chem.* 153: 375-380.

Reyes, V.G. & J.M. Tiedje 1976. Ecology of the gut microbiota of *Tracheoniscus rathkei* (Crustacea, Isopoda). *Pedobiologia* 16: 67-74.

Rushton, S.P. & M. Hassall 1983. Food and feeding rates of the terrestrial isopod *Armadillidium vulgare* (Latreille). *Oecologia* 57: 415-419.

Scriber, J.M. & F. Slansky Jr 1981. The nutritional ecology of immature insects. *Annu. Rev. Entomol.* 26: 183-211.

Shachak, M., E.A. Chapman, & Y. Steinberger 1976. Feeding, energy flow and soil turnover in the desert isopod, *Hemilepistus reaumuri. Oecologia* 24: 57-69.

Slansky Jr, F. & G.S. Wheeler 1989. Compensatory increases in food consumption and utilization efficiencies by velvetbean caterpillars mitigate impact of diluted diets on growth. *Entomol. exp. appl.* 51: 175-187.

Stafford, H.A. & H.H. Lester 1986. Proanthocyanidins in needles from six genera of the Taxodiaceae. *Amer. J. Bot.* 73: 1555-1562.

Tabashnik, B.E. & F. Slansky 1987. Nutritional ecology of forb foliage-chewing insects In: Slansky F, Rodriguez J G (eds.) *Nutritional ecology of insects, mites, spiders, and related invertebrates.*, pp 71-103. Wiley & Sons, New York.

Teuben, A. 1991. Nutrient availability and interactions between soil arthropods and microorganisms during decomposition of coniferous litter: a mesocosm study. *Biol. Fertil. Soils* 10: 256-266.

Teuben, A. & T.A.P.J. Roelofsma 1990. Dynamic interactions between functional groups of soil arthropods and microorganisms during decomposition of coniferous litter in microcosm experiments. *Biol. Fertil. Soils* 9: 145-151.

Thalmann, A. 1968. Zur Methodik der Bestimmung der Dehydrogenase-Aktivität im Boden mittels Triphenyltetrazoliumchlorid (TTC). *Landwirtsch. Forsch.* 21: 249-258.

Topp, W. & P. Beracz 1989. Effect of host plant and changing seasonal development on consumption rates, utilization efficiencies and survival of *Melasoma punctata* (Scop.) (Col., Chrysomelidae). *J. Appl. Ent.* 107: 261-274.

Ullrich, B., V. Storch, & H. Schairer 1991. Bacteria on the food, in the intestine and on the faeces of the woodlouse *Oniscus asellus* (Crustacea, Isopoda). *Pedobiologia* 35: 41-51.

Waldbauer, G.P. 1968. The consumption and utilization of food by insects. *Adv. Insect. Physiol.* 5: 229-288.

Warburg, M.R. 1987. Isopods and their terrestrial environment. *Adv. Ecol. Res.* 17: 187-242.

Wieser, W. 1965. Untersuchungen über die Ernährung und den Gesamtstoffwechsel von *Porcellio scaber* (Crustacea: Isopoda). *Pedobiologia* 5: 304-331.

Microhabitat selection by *Armadillidium vulgare* in a riparian forest: Lack of apparent influence by leaf litter food quality

F. Heinzelmann, C.S. Crawford & M.C. Molles, Jr.
Department of Biology, University of New Mexico, Albuquerque, USA

M.R. Warburg
Department of Biology, Technion-Israel Institute of Technology, Haifa, Israel

ABSTRACT

Armadillidium vulgare is the only isopod in the riparian forest (or 'bosque') at the Rio Grande Nature Center in Albuquerque, New Mexico. The forest there is dominated by native cottonwood, *Populus fremontii* var. *wislizeni*, and Russian olive, *Eleagnus angustifolia*. We tested whether leaf litter of these two tree species influences the isopod's microhabitat selection in this relatively simple setting by pitfall trapping during the warm season over a period of 4 years. We evaluated the quality of leaf litter as food by studying growth and survival of *Armadillidium* fed cottonwood and/or Russian olive leaf litter. ANOVAs of field results indicated that tree species did not influence the distribution of isopods beneath the canopies of four pairs of trees of each species. However, significantly larger catches occurred under trees, regardless of their species, located in clay loam soils than under trees located in sandy loam soils. ANOVAs of laboratory feeding tests showed that *Armadillidium* grew and survived significantly better on Russian olive and mixed species leaf litter than it did on cottonwood leaf litter. We conclude that the distribution of leaf litter under the two tree species at the Nature Center does not influence microhabitat selection by *Armadillidium*.

1 INTRODUCTION

Oniscoid isopods transplanted from Europe and the Mediterranean fringe to the New World are represented by at least a dozen species in mesic parts of North America; however, only a few, widespread species have colonized the comparatively arid U.S. Southwest (Muchmore 1990). These now occur in a variety of habitats (e.g. riparian forests, urban gardens) that generally provide appropriate combinations of food and moisture. Within such habitats the distribution of these isopods tends to be patchy, indicating they select microhabitats where certain conditions are optimal for a given period. Some microhabitats, such as spaces beneath logs, are obvious; others, such as specific locations in leaf litter, are not. Little is known about microhabitat selection by introduced isopods in this region.

In general, the occupation of a habitat by terrestrial isopods is influenced by substrate, moisture, shelter, and climate (see discussions in Warburg et al. 1984; Warburg 1993). Food, in contrast, appears to be of limited importance in this regard (Warburg et al. 1984; Hornung et al. 1991). Yet terrestrial isopods do exhibit feeding preferences for different types of leaf litter (e.g. Hassall & Rushton 1984; Dudgeon et al. 1990; Szlávecz & Maiorana 1991), and they also demonstrate differing growth and survival performances on different types of food (Rushton & Hassall 1983). Therefore, the possibility that their microhabitat preferences reflect spatial and temporal patterns of leaf litter distribution and quality cannot be ruled out.

This study addresses isopod microhabitat selection in a relatively simple natural setting: a riparian forest dominated by two tree species and colonized by one species of Old World isopod. In this report we first describe the distribution of the isopod beneath individual trees over a 4-y period. We then document its capacity to grow and survive on leaf litter of the two dominant trees. This combination of field and laboratory research provides us with data to test, for this species under these conditions, whether leaf litter is an important factor in its selection of microhabitat.

2 MATERIALS AND METHODS

2.1 *Study site description and pitfall trap monitoring*

The Rio Grande Nature Center in Albuquerque, New Mexico consists mainly of a riparian forest (regionally known as the 'bosque'). The dominant forest trees are mature Rio Grande cottonwood, *Populus fremontii* var. *wislizeni* Watts and mixed-age Russian olive, *Eleagnus angustifolia* L. The bosque also contains small numbers of native 'tree willows,' e.g. *Salix goodingii* Pall., as well as introduced Siberian elm, *Ulmus pumilla* L., and salt cedar, *Tamarix pentandra* Ball. Its sparse woody understory is characterized mainly by native coyote willows, *Salix exigua* Nutt., and clumps of false indigobush, *Amorpha fruticosa* L., a legume. The bosque's herbaceous understory is weakly developed, especially where leaf litter has accumulated beneath trees. The legumes *Melilotus albus* Descr. and *Sphaerophysa salsula* (Pall.) DC., as well as the composites *Machaeranthera* spp. are moderately abundant in relatively open spaces during years of above average precipitation. Small amounts of grass also occur in unshaded spaces. The mustard *Descurainia pinnata* (Walt.) is present under the canopies of some Russian olive trees following relatively wet winters.

The pill bug, *Armadillidium vulgare* Latrielle, is the only terrestrial isopod in the Nature Center's bosque, where its activity, together with that of other epigeic arthropods, is monitored with pitfall traps. Traps were made of plastic cups, 9 cm in diameter at the opening and 11.5 cm deep. Two cups, one inside the other, were positioned at the ground surface and covered with a 15 × 15 cm plywood square. Screws in the corners allowed the cover to be raised when a trap was in use.

In this study, four traps, each situated in a cardinal direction, were placed halfway between the trunk and canopy edge of selected trees. This trap configuration surrounded four designated pairs of cottonwood and Russian olive trees within a north-south forest strip approximately 900 m in length (Fig. 1). Individual trees in a pair

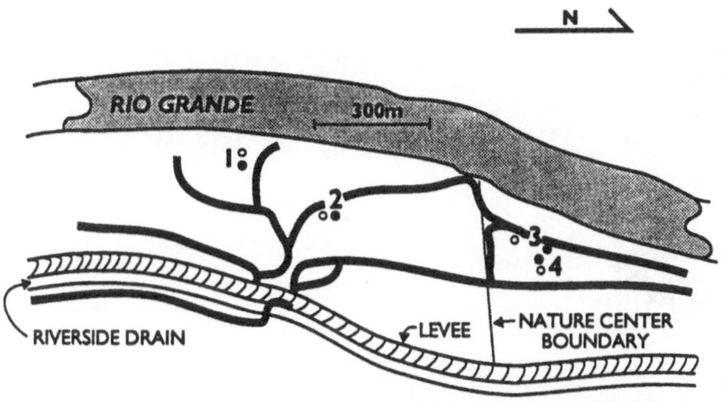

Figure 1. Diagram of the northern Rio Grande Nature Center, showing approximate locations of members of the four pairs of cottonwood (●) and Russian olive (○) trees used in this study. Trails (black lines) and other landscape features are shown for reference. Soils in the vicinity of pairs 1 and 2 are clay loam; soils in the vicinity of pairs 3 and 4 are sandy loam.

were 10-60 m apart. They were chosen for their relative isolation from trees of the other dominant species in this mixed-species forest. A gradient of soil texture exists in the strip; clay loams characterize the south end and sandy loams characterize the north end. For purposes of analysis we therefore divided the strip into a south block and a north block.

Pitfall traps beneath the four pairs of trees were opened for 48 h at 2-5 wk intervals between September 1989 and September 1993. Trap contents from beneath each tree were pooled and frozen for later examination.

2.2 *Laboratory studies of Armadillidium growth and survival*

Laboratory studies involved two approaches to test whether growth and survival were superior on leaf litter of cottonwood or leaf litter of Russian olive. The first approach used offspring of isolated females fed only one of the two types of leaf litter. The second approach, begun over a year later, used offspring of females maintained in groups on different leaf litter treatments. The rearing and experimental protocols of both approaches are outlined in Figures 2 and 3.

In the first approach (Fig. 2), we collected adult females (assumed mated) on 20 March 1991 from leaf litter and beneath fallen wood at the Nature Center. Cottonwood and Russian olive were both common at the collection site. Isopods were kept individually, in a laboratory, in clear plastic coffee creamers (height: 4.5 cm, diameter: 3.5 cm) with tight-fitting lids. Clay-loam habitat soil, collected to a depth of 4 cm from the Nature Center, was added to each creamer to a depth of 1.5 cm. Twenty of the isopods were given Russian olive leaf litter and 20 others were given cottonwood litter. Litter, and water administered by pipette, were replenished every 10 d. Laboratory temperature was 20-22°C. Creamers were shaded but photoperiod was not controlled.

Leaf litter used as food for all of the first approach and part of the second was taken from baskets stored on the ground and made from aluminum window screening. The baskets (height: 10 cm, width: 0.6 m) had screen covers and contained

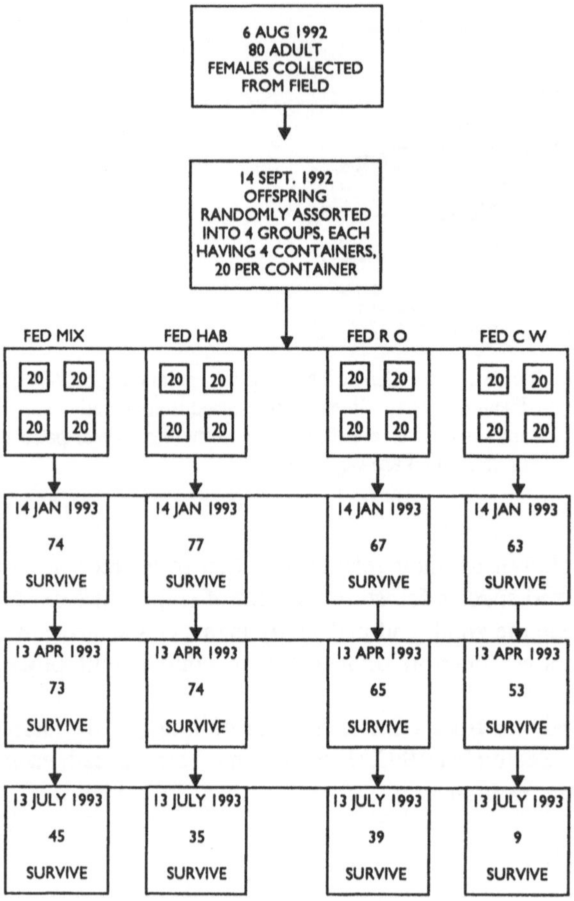

Figure 2. Feeding and rearing flow chart for offspring of female *Armadillidium* collected from Rio Grande Nature Center on 20 March 1991. CW refers to cottonwood leaf litter; RO refers to Russian olive leaf litter.

leaves from one tree species or the other. They were kept on the ground under a Russian olive-cottonwood canopy at the Nature Center. Leaves in the baskets were first collected in October 1990, just prior to falling from trees in various Nature Center locations. Leaves used for food were initially taken from the baskets in February 1991 and frozen until needed.

On 17 January 1992, we removed the 224 offspring born between the previous July and September to the seven gravid females in the cottonwood-fed group. (Since only 68 offspring of the females fed Russian-olive had been produced by that time, we used just the descendants of the cottonwood-fed females in the next phase of this approach). The numbers, per adult female, of these offspring ranged from 15 to 45. Twenty-one of the small isopods were randomly transferred to each of eight translucent square plastic containers (height: 7.5 cm, width: 10 cm) with screened openings in their lids. Soil as described above was added to the containers to a depth of 4 cm. Water was administered by pipette every 5 d.

Four containers received Russian olive leaf litter and four received cottonwood leaf litter; the litter had been collected in October 1991 and was taken from the field baskets at 2 wk intervals shortly before use. Leaves were refrigerated but not frozen.

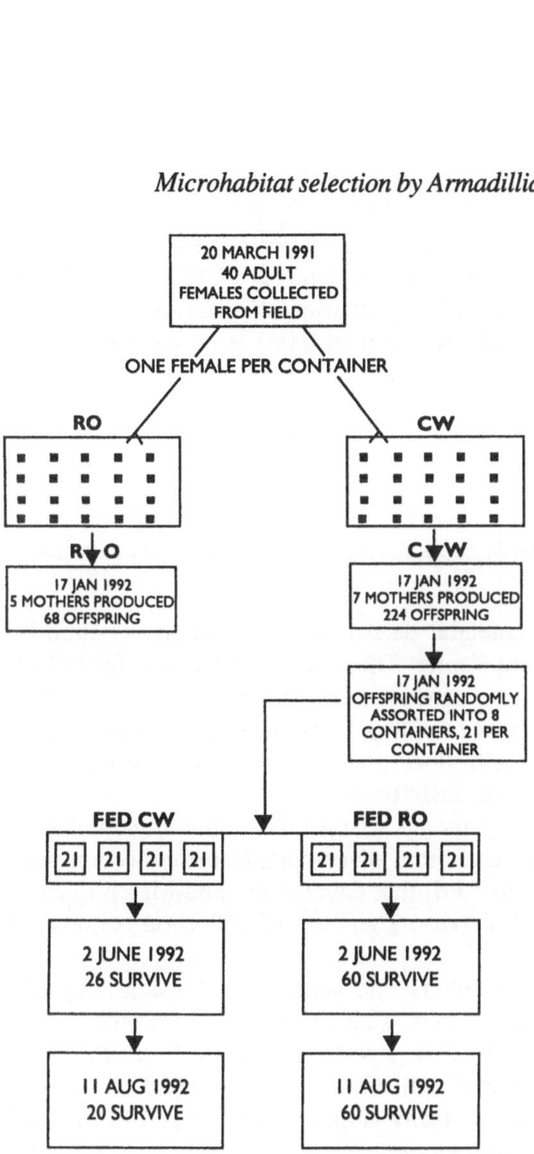

Figure 3. Feeding and rearing flow chart for offspring of female *Armadillidium* collected from Rio Grande Nature Center on 6 August 1992. CW refers to cottonwood leaf litter; RO refers to Russian olive leaf litter; MIX refers to an artificial mixture of both leaf litter types; HAB refers to leaf litter (essentially the same mixture) collected from the habitat of the original females.

The isopod containers were kept at 20-22°C in a chamber with photoperiod controls that approximated seasonal daylengths. The light source was a fluorescent Vita Lite full spectrum bulb situated approximately 50 cm above the containers. This temperature and photoperiod arrangement was used in all subsequent rearing.

The second approach (Fig. 3) was begun on 6 August 1992, when 80 females (again assumed mated) were collected from a site at the south end of the soil texture gradient mentioned above. Equal numbers were randomly assorted into four of the above-described translucent containers. Soil from the collection site was added to each container; leaf litter from both tree species, collected randomly from the site and moistened periodically, was placed on top of the soil. Offspring were soon observed in the containers; on 14 September we randomly selected 320 of them and added 20 of each to clay-loam soil in 16 similar containers.

These containers were divided into four groups, each containing four replicates. Each group received a different leaf litter treatment. Fresh supplies of litter were added to each container at 3 wk intervals; equal amounts of water were added to each container every 5 d. One set ('HAB') received leaf litter gathered randomly from the capture site habitat. The three other sets received leaves added to the field baskets in

October 1992 and handled as described above for the previous year. One of these ('MIX') received a mixture of the two types of leaves. Another ('CW') received only cottonwood leaves. The third ('RO') received only Russian olive leaves. All isopods used in this experiment were counted and weighed to 0.001 g on 14 January, 13 April, and 13 July 1993.

3 RESULTS

3.1 *Habitat selection by Armadillidium beneath individual cottonwood and Russian olive trees*

We considered three variables to be potential determinants of habitat selection by *Armadillidium* at the Rio Grande Nature Center. These were (1) tree species (which directly influenced the type of leaf litter beneath their canopies), (2) soil texture (which in this instance reflected the eight sampling locations under cottonwood or Russian olive, four in the clay-loam south block and four in the sandy-loam north block), and (3) year (which reflected annual differences in climate).

In order to standardize the influence of these variables on *Armadillidium* catches over a span of four years, we only used data from seven comparable collection periods (each separated among years by 8 to 15 d) that covered the mid-March to early September timing of significant isopod activity. The dates of collection periods are given in Table 1.

ANOVA (Table 2) suggests that soil texture and year, but not species, significantly influenced *Armadillidium* activity. Figure 4 also shows that tree species had no apparent influence on average annual catch ($P = 0.827$). Figure 5, on the other hand, shows that there were apparent significant effects by soil texture on average annual catch ($P = 0.008$). Roughly twice as many isopods were trapped in the clay loam south block as in the sandy loam north block.

In addition, Figure 5 indicates that year was significantly associated with average total catch per tree ($P = 0.008$). The possible influence of yearly antecedent precipitation was evaluated using Nature Center weather records, which gave the following totals from December through April: 118 mm in 1989-1990, 50 mm in 1990-1991, 84 mm in 1991-1992, and 128 mm in 1992-1993. Regression of annual catch totals of *Armadillidium* on the log-transformed precipitation totals is not statistically sig-

Table 1. Among-year comparable pitfall trap collection dates used to analyze the influence of tree species, soil texture, and year on *Armadillidium* habitat selection.

1990	1991	1992	1993
23 March	22 March	27 March	12 March
6 April	5 April	10 April	2 April
20 April	19 April	24 April	30 April
18 May	17 May	21 May	26 May
22 June	28 June	19 June	23 June
13 July	19 July	17 July	26 July
24 August	30 August	27 August	3 September

Table 2. Multiple ANOVA comparing the effects of tree species, soil texture (in north & south study area 'blocks'), and year on *Armadillidium* pitfall catches at the Rio Grande Nature Center.

Source	SS	df	MS	F	P
species	0.855	1	0.855	0.049	0.827
block	163.835	1	163.835	9.470	0.008
year	294.697	3	98.232	5.678	0.008
species × block	25.257	1	25.257	1.460	0.246
block × year	40.977	3	13.659	0.790	0.518
species × year	5.237	3	1.756	0.101	0.958
species × block × year	3.980	3	1.327	0.077	0.972
error	259.505	15	17.300		

Multiple $R^2 = 0.694$

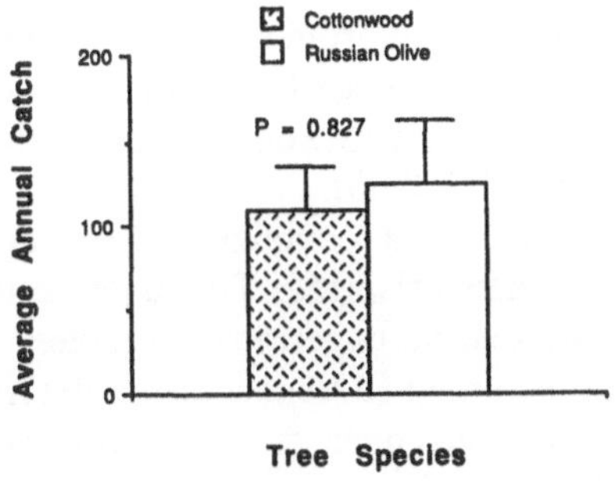

Figure 4. Lack of apparent influence of the two dominant tree species at Rio Grande Nature Center on average annual pitfall catch of *Armadillidium* over a 4-y period.

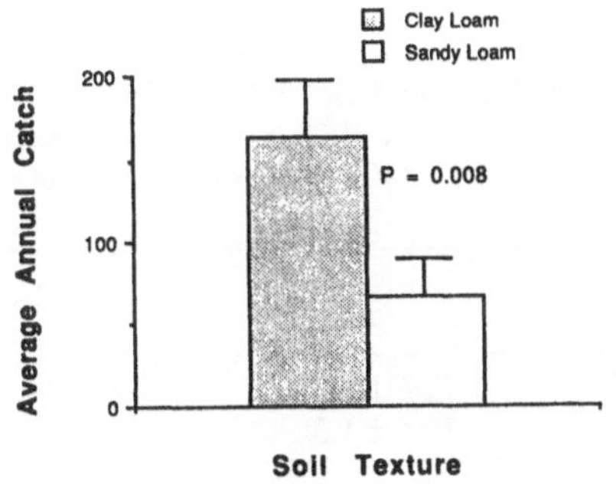

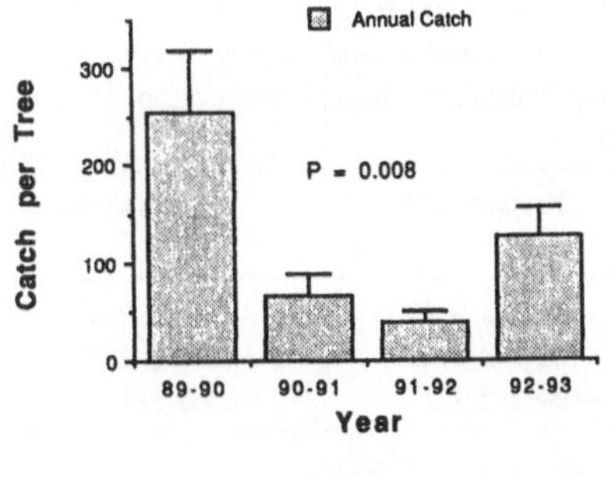

Figure 5. Top: Apparent influence along a gradient of soil texture (clay loam at the south end and sandy loam at the north end) on average annual catch of *Armadillidium* at Rio Grande Nature Center. Bottom: Apparent influence of year on average annual catch of *Armadillidium* per tree (among the four study pairs of cottonwood and Russian olive trees at Rio Grande Nature Center).

nificant (P > 0.1); however, the resulting r-square of 0.45 suggests that winter-spring precipitation may be shown to affect the variability in average total catch per tree when more data become available.

3.2 *Growth and survival of Armadillidium on cottonwood and Russian olive leaf litter*

Results of the first approach to testing whether growth and survival were superior on leaf litter of cottonwood or Russian olive are given in Figure 2 and Table 3. Significantly fewer offspring of cottonwood-fed, individually maintained females survived on cottonwood leaf litter than on Russian olive leaf litter (ANOVA, P < 0.05). The cottonwood litter juveniles also weighed significantly less than the Russian olive litter juveniles after five (and seven) months (ANOVA, P < 0.05).

The same general result is seen in Figure 3 and Table 4, which compare the growth and survival of offspring born to females taken directly from the field and fed on field-collected leaf litter. Those juvenile *Armadillidium* reared only on cottonwood survived in progressively fewer numbers and continued to weigh significantly less than those reared on the other three diets (ANOVA, P < 0.05).

Maintaining cottonwood-fed females in small individual containers (first approach) may have adversely affected the growth of their offspring. This is evident when the last (summer) set of weights from the two rearing experiments (Tables 3 and 4) are compared: weights of juveniles in all non-cottonwood treatments shown in Table 4 are roughly 1.5 times greater than weights of juveniles fed Russian olive (Table 3). The same relationship holds for juveniles fed only cottonwood in the two experiments.

Table 3. Live weights (mg) of surviving *Armadillidium* offspring from field-collected females fed cottonwood leaf litter, then transferred in equal numbers on 17 January 1992 to four replicates of cottonwood and Russian olive leaf litter treatments. Compared by ANOVA followed by Fisher's multiple comparison.

1992 Date	Russian olive $\bar{x}$ ±SE		P	Cottonwood $\bar{x}$ ±SE
2 June	26 ± 1.2	<	0.05	12.4 ± 0.8
11 August	38.7 ± 1.4	<	0.05	17.4 ± 1.3

Table 4. Live weights (mg) of surviving *Armadillidium* offspring from field-collected females fed habitat litter, then transferred in equal numbers on 14 September 1992 to four replicates of the following leaf litter treatments: mixed, habitat, Russian olive, and cottonwood. See text and Figure 3 for further explanation. Compared by ANOVA followed by Fisher's multiple comparison. Different letters in a row indicate significant differences at P = 0.05.

1993 Date	Mixed $\bar{x}$ ±SE	Habitat $\bar{x}$ ±SE	Russian olive $\bar{x}$ ±SE	Cottonwood $\bar{x}$ ±SE
14 January	43.9 ± 4.3[a]	41.4 ± 10.4[a]	40.2 ± 9.4[a]	16.3 ± 6.2[b]
13 April	62.6 ± 5.8[a]	50.0 ± 9.9[a]	60.5 ± 13.3[a]	19.0 ± 4.7[b]
13 July	59.9 ± 1.7[a]	65.7 ± 2.2[a]	54.1 ± 1.7[a]	24.7 ± 1.3[b]

[a] Means in columns not followed by the same letter are significantly different.

4 DISCUSSION

The capacity of *Armadillidium vulgare* to grow and survive on a variety of diets (Paris & Sikora 1967; Watanabe 1978) must facilitate its broad distribution in the Rio Grande Nature Center's riparian forest. Our results, however, imply that the distribution of its most apparent food resource, leaf litter of cottonwood and Russian olive, does not determine its distribution on a smaller scale. More likely, microhabitat selection by *Armadillidium* in the Nature Center is a function of soil texture and the overall influence of yearly climate on its immediate environment. Other factors, such as topography of the microhabitat, extent of woody litter accumulation, colonization of leaf litter by microorganisms, and presence of rodent burrows may also be involved. Although the availability of living plants as food (Beck & Bretowsky 1980; Szlávecz & Maiorana 1990) cannot be discounted, we have not seen isopods feeding on green vegetation beneath trees in this forest.

Results not unlike ours are seldom reported for terrestrial isopods. However, a somewhat similar case is that of *Trachelipus rathkei* Brdt., one of four species studied by Szlávecz (1992) on a lake shore in Hungary. *T. rathkei* was evenly distributed on detrital drifts of submerged and emerged aquatic vegetation, these being of high and low nutrient quality, respectively. It consumed both in equal amounts even though it and the other isopods lost weight when fed the emerged detritus but gained weight when fed its counterpart. The much rarer *A. vulgare* consumed more of the formerly submerged detritus, but only occurred on the emergent detritus; thus it, too, may have demonstrated the type of microhabitat selection shown by the Nature Center population.

While small scale patterns of an isopod's spatial distribution over time may not depend always on the actual nutritional effects of leaf litter, such material can be important in other ways. For example, work by Merriam (1971) illustrated that food quality has distinct effects on individual growth rates of *A. vulgare*, and that these rates, in turn, affect natality because timing of first reproduction depends on size of female parent.

We have shown that a diet of (introduced) Russian olive leaf litter, which contains significantly more nitrogen than (native) cottonwood leaf litter (M. Parker, unpublished data), results in comparatively rapid growth and strong survival in Nature Center *Armadillidium.* However, the differential effects of the two leaf litter types on reproduction in this population are less clear and require further investigation. In our study, the cottonwood diet fed to individually housed females was associated with greater realized fecundity (and lower survival) than the Russian olive diet fed to other isolated females. At least two factors could have been involved. First, the cottonwood diet may have so stressed the females that they released as many offspring as possible before they died. Also, the original sizes of the females could have been a factor in their survival (Lawlor 1976; Brody et al. 1983). We think not, in this case, because regressions of female weights, taken within a month of parturition, on numbers of offspring produced by each female, have very low r-square values.

Several other studies have examined the distribution of isopods under trees. Howard (1980) found that the great majority of pregnant female *A. vulgare* under a single tree congregated close to the trunk on the most humid side. Earlier, Brereton (1957) recorded 'micro-geographical separation' of four species of isopods among trees in an oak-ash-sycamore wood in England; interestingly, *A. vulgare* was rare inside the wood but common under stones just outside.

The spatial and temporal distribution of *A. vulgare* in the Rio Grande bosque awaits a more complete explanation. Our findings, however, support the hypothesis that leaf litter quality, as it affects the isopod's growth and survival, is not an important factor in its selection of microhabitat.

ACKNOWLEDGEMENTS

We are indebted to the many 'Bosque Biology' students at the University of New Mexico who assisted in isopod data collection, and to the Rio Grande Nature Center personnel who gave us support. We also thank Beth Dennis, Katelijne Flies, and Melissa Parker for helping in other ways. We appreciate comments on an earlier version of the manuscript by Lisa Ellis and an unknown reviewer. Katelin Szlávecz provided a very useful review of the final version.

REFERENCES

Beck, L. & E. Brestowsky 1980. Auswahl und Ververtung verschiedener Fallaubarten durch *Oniscus asellus* (Isopoda). *Pedobiol.* 20: 428-441.

Brereton, J.L.G. 1957. The distribution of woodland isopods. *Oikos* 8:85-106.

Brody, M.S., Edgar, M.H. & L.R. Lawlor 1983. A cost of reproduction in a terrestrial isopod. *Evolution* 37: 653-655.

Dudgeon, D., Ma, H.H.T. & P.K.S. Lam 1990. Differential palatability of leaf litter to four sympatric isopods in a Hong Kong forest. *Oecologia* 84: 398-403.

Hassall, M. & S.P. Rushton 1984. Feeding behaviour of terrestrial isopods in relation to plant defences and microbial activity. In S.L. Sutton & D.L. Holdich (eds.), *The Biology of Terrestrial Isopods. Symposia of the Zoological Society of London No. 53*, pp. 487-505. Oxford: Clarendon Press.

Hornung, E., Warburg, M.R. & K. Szlávecz. 1991. Trends and methods in terrestrial isopod ecology: a round-table discussion. *Proceedings of the Fourth European Congress of Entomology. SIEEC*, Godollo: 747-750.

Howard, H.W. 1980. The distribution at breeding time of the sexes of the woodlouse, *Armadillidium vulgare. Crustaceana* 39: 52-56.

Lawlor, L.R. 1976. Molting, growth and reproductive strategies in the terrestrial isopod, *Armadillidium vulgare. Evolution* 57:1179-1194.

Merriam, H.G. 1971. Sensitivity of terrestrial isopod populations (*Armadillidium*) to food quality differences. *Can. J. Zool.* 49: 667-674).

Muchmore, W.B. 1990. Terrestrial Isopoda. In D.L Dindal (ed.), *Soil Biology Guide*, pp 805-817. New York: John Wiley & Sons, Inc.

Paris, O.H. & A. Sikora. 1967. Radiotracer analysis of the trophic dynamics of natural isopod populations. In K. Pertrusewicz (ed.), *Secondary Productivity of Terrestrial Ecosystems (Principles and Methods), vol. II*, pp 741-771. Warsaw: Institute of Polish Academy of Sciences.

Rushton, S.P. & M. Hassall. 1983. The effects of food quality on the life history parameters of the terrestrial isopod *Armadillidium vulgare* (Latrielle). *Oecologia* 57: 257-261.

Szlávecz, K. 1992. The role of terrestrial isopods (Isopoda: Oniscidea) in the decomposition of aquatic macrophyte detritus of Lake Balaton, Hungary. *Opusc. Zool. Budapest* 25: 103-112.

Szlávecz, K. & V.C. Maiorana. 1991. Food selection by isopods in paired choice tests. In P.

Juchault & M. Moquard (Eds.), *Biology of Terrestrial Isopods, Third International Symposium.* Poitiers, pp. 115-119. Poitiers: University Press.

Warburg, M.R. 1993. *Evolutionary Biology of Land Arthropods*. Berlin: Springer-Verlag.

Warburg, M.R., Linsenmair, K. & Berkovitz. 1984. The effect of climate on the distribution and abundance of isopods. In S.L. Sutton & D.L. Holdich (eds.), *The Biology of Terrestrial Isopods. Sypmposia of the Zoological Society of London No. 53*, pp. 339-367. Oxford: Clarendon Press.

Watanabe, H. 1978. A food selection experiment on palatability of woodlouse, *Armadillidium vulgare. Edaphologia* 18: 2-8.

Population genetics of *Armadillidium vulgare* in Europe and North America

R. L. Garthwaite
Institute of Marine Sciences, University of California, Santa Cruz, Calif., USA; and Osher Foundation Laboratory for Molecular Systematics, California Academy of Sciences, San Francisco, Calif., USA

R. Lawson
Osher Foundation Laboratory for Molecular Systematics, California Academy of Sciences, San Francisco, Calif., USA; and Leeds City Museum, Natural History Dept., Calverly St., Leeds, UK

C. Sassaman
Department of Biology, University of California, Riverside, Calif., USA

ABSTRACT

A survey of genetic variation at up to nine enzyme loci in the synanthropic oniscid isopod *Armadillidium vulgare* (Latreille, 1804) was conducted from 1980 to 1992 using 157 population samples comprising over 10,000 individuals from throughout Europe and North America. Based on analysis of F-statistics, most observed genetic variation in this species in both Europe and North America was found within populations. In Europe, allele frequencies were generally found to be highly variable among populations with large differences occurring over relatively short distances. Northern Europe was found to be substantially different from southern Europe. In contrast, in North America, allele frequencies were found to be more broadly similar with smaller differences among regions. Similarities of allele frequencies among regions of North America and Europe, as measured by F-statistics and allele distributions, were found to correlate with expected similarities based on the history of human colonization of North America by Europeans. However, due to the similarity in climatic and environmental factors between likely European source areas and areas of potential colonization in North America, the observed similarities in allele frequencies would also support a selective interpretation of the data. The patterning of genetic variation in *A. vulgare* in Europe and North America is discussed in light of selective and historic theories.

1 INTRODUCTION

It is a primary goal of population geneticists and evolutionary biologists to delineate the factors which effect genes in populations and to determine how these factors contribute to evolutionary change. Unfortunately, evolutionarily important factors often, though not always, have important effects over time periods far longer than scientific careers. For this reason, observations of single populations over time rarely contribute substantially to our understanding of evolution. An alternate approach is the study of spatial variation. Spatially separated populations are a reasonable substitute for temporally separated generations since spatially separated populations are separated in time from a common ancestor (Singh & Long 1992). The nature of genetic variation among populations can tell us about the mechanism of evolutionary change. The importance of spatial variation in evolutionary studies is obvious and has been a major area of investigation since the early days of the evolutionary synthesis (Singh & Long 1992; see Darwin 1859).

Oniscid isopods possess a number of biological characteristics which make them particularly interesting subjects for evolutionary studies. These characters include multiple mating and the production of broods with multiple paternity (Howard 1943; Sassaman 1978b), sperm retention and the production of multiple broods from a single mating (Vandel 1941; Howard 1943; Hatchett 1947; Lueken 1962), high fecundity with up to over 200 offspring per brood possible (Collinge 1915), and tolerances for wide ranges of geographic, altitudinal, and environmental variation.

Due in part to these characteristics and its synanthropic nature, the oniscid isopod *Armadillidium vulgare* (Latreille 1804) is a very successful colonizer. From its origins in southern Europe (Vandel 1962), *A. vulgare* has spread into terrestrial habitats over much of the temperate and subtropical areas of the earth. Europe and the Mediterranean (Sars 1899, Vandel 1962, this paper), the Arabian Peninsula (Taiti & Ferrara 1991), much of North and South America (Van Name 1936, 1942; Mulaik 1960; this paper), southern Africa (Ferrara & Taiti 1979), and Australia (Green 1961) have been colonized as well as New Zealand (Chilton 1901; Jackson 1941), Japan and China (Silvestri 1927; Saito 1986), and many smaller Pacific and Atlantic islands (Van Name 1942; Vandel 1962; Ferrara & Taiti 1979, this paper). In the contiguous United States *A. vulgare* is ubiquitous with populations ranging from high elevations, e.g. Denver, Colorado and Santa Fe, New Mexico, to the subtropics south of Miami, Florida.

Because *Armadillidium vulgare* has been introduced by man into a wide range of environments from its ancestral home in southern Europe, it represents an excellent model system to test theories concerning the adaptive importance of genetic variation. By sampling genetic variation in this species over a wide range of non-native localities, we can observe its genetic response to new and vastly different environmental conditions. These observations should give insights into the role of genetic variation in adaptation to the environment. For example, correlations of genetic variation in new, non-native localities with the source of colonization regardless of climatic or environmental factors would indicate that historical rather than selective factors dictate the patterning of genetic variation in this species, whereas correlations of genetic variation with new climatic or environmental factors regardless of source of introduction would argue for genetic adaptation to these new environments.

In this initial paper on genetic variation in *Armadillidium vulgare*, we document patterns of allele frequencies at nine polymorphic enzyme-coding loci in natural populations from collection sites throughout Europe and North America. This data is used to determine whether the pattern of genetic variation in North American populations of *A. vulgare* correlates with known patterns of settlement of North America by European peoples, or whether there is evidence that this genetic variation is the result of adaptation to a variety of terrestrial habitats.

2 MATERIALS AND METHODS

Horizontal starch gel electrophoresis was used to determine allele frequencies at up to nine loci coding for soluble proteins in 157 population samples of *Armadillidium vulgare* collected from 1980 to 1992 and comprising a total of over 10,000 individuals from throughout Europe and North America. Samples were usually collected from areas not exceeding $2m^2$ and for the purposes of this study we consider each of these to represent a deme. These samples were transported either alive or frozen to our laboratories in California where they were maintained as live cultures or kept frozen at $-80^{\circ}C$ until required for electrophoresis. Collection localities for these samples are listed in Appendix 1. Collection sites are shown in Fig. 1.

Tissue extracts were prepared for electrophoresis by homogenizing whole isopods or one or more isopod legs in 0.05M Tris–HCl pH 7.5, or in 0.25M sucrose, 10mM dithiothreitol, 10mM EDTA, 10mM $MgCl_2$, and 0.01M Tris adjusted to pH 7.5 with HCl. Approximately two volumes of homogenizing buffer were used for each volume of isopod tissue. The enzyme loci surveyed and electrophoretic procedures are summarized in Table 1.

Allelomorphs were detected using methods from Murphy et al. (1990) and Aebersold et al. (1987) with minor modifications. Most population samples (143) were assayed for variation at the AAT–1, IDH–1, MDH–1, MDH–2, GPI, and PGM loci, but for MPI, IDH–2, and PEPGL a smaller number of populations was assayed.

For some data analyses, the populations were grouped into six regions. Europe was divided into two regions (north Europe and south Europe) at 46° north latitude (Fig. 1). North America was divided into four regions (northeast, southeast, northwest, and southwest) at $36^{\circ}30'$ north latitude and 98° longitude (Fig. 1).

For two loci (GPI and PGM), the relationship between the frequency of the most common allele and latitude were measured using the Pearson product-moment correlation coefficient (Sokal & Rohlf 1981). Resulting coefficients were tested for significance using a table of critical values for correlation coefficients (Rohlf & Sokal 1981).

Average number of alleles per locus, mean observed heterozygosity by direct count, and expected heterozygosity (unbiased estimate based on conditional Hardy-Weinberg expectations; Levene, 1949; Nei, 1978) were calculated. For comparison, average number of alleles per locus and mean observed heterozygosity were averaged over samples within regions.

Genotype frequencies were tested for conformity to Hardy-Weinberg equilibria with the usual chi-square goodness-of-fit test using observed genotype frequencies and those expected under Hardy-Weinberg equilibrium, and Cooper's (1968) correction for multiple tests. Expected frequencies were calculated using Levene's (1949) formula for small sample sizes.

Additionally, overall within-region genetic equilibrium was assessed using the inbreeding coefficient F_{is} (Nei 1977) averaged over alleles. For each locus we tested the null hypothesis of $F_{is} = 0$ using $X^2 = n_t(F_{is})^2$ where n_t is the total of all individuals in all population samples (Li & Horvitz 1953; Nei & Chesser 1983). In the case of multiple populations, the degrees of freedom are (k–1) (s–1) where k is the number of alleles at a locus and s is the number of populations sampled.

In considering differentiation within and between regions we used the hierarchical

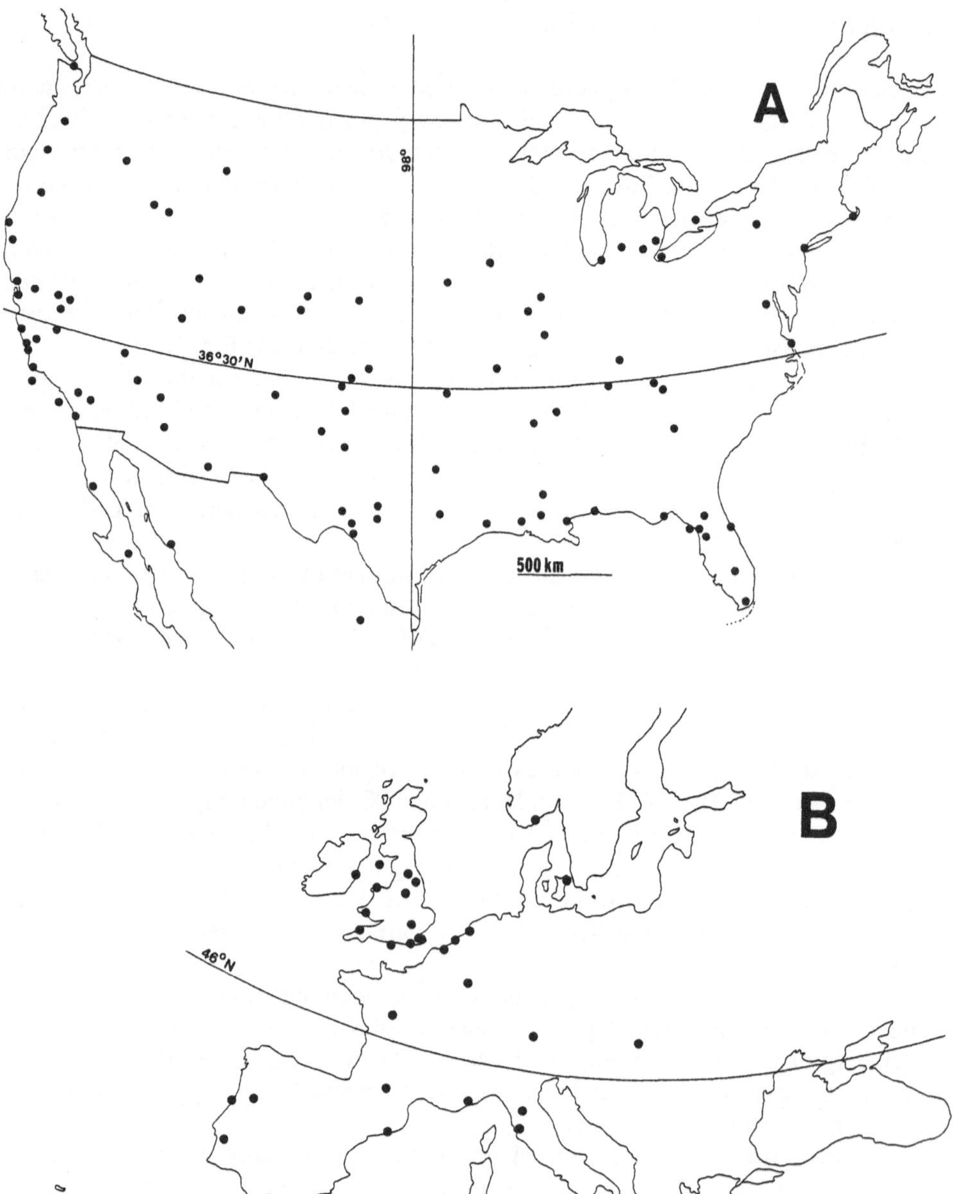

Figure 1. Location of *Armadillidium vulgare* sample collection sites in A) North America and B) Europe. Each dot represents one or more collection sites.

Table 1. Summary of electrophoretic procedures including the loci scored, the number of alleles detected, and the buffer systems used.

Enzyme	I.U.B. number	Loci scored	Number of alleles detected	Buffer systems*
Aspartate aminotransferase	2.6.1.1	AAT–1	4	1,2
Glucose-6-phosphate isomerase	5.3.1.9	GPI	6	2,3,4
Isocitrate dehydrogenase	1.1.1.42	IDH–1	5	1,2,4
		IDH–2	2	1,2,4
Malate dehydrogenase	1.1.1.37	MDH–1	4	1,2
		MDH–2	2	1,2
Mannose-6-phosphate isomerase	5.3.1.8	MPI	5	1,2
Dipeptidase (gly-leu)	3.4.?.?	PEPGL	4	2,3
Phosphoglucomutase	5.4.2.2	PGM	4	2,3,4

*Buffer systems: 1) = buffer system 5 of Selander et al. (1971). 2) = (amino-propyl)-morpholine system of Clayton & Tretiak (1972) at pH 7.5 or 8.0. 3) = buffer system 2 of Selander et al. (1971). 4) = buffer system 9 of Selander et al. (1971).

F-statistics of Wright (1978). The allele frequency data was partitioned at three levels for all possible pair-wise comparisons among regions. These are the population (deme), the region, and the total. For each pair of regions there will then be three F-statistics analogous to the F_{st}. These are for populations within regions, F_{pr}, populations within the total, F_{pt}, and regions within the total, F_{rt}. Pair-wise values of F_{rt} were clustered using the unweighted pair-group method, arithmetic averages, UPGMA (Sneath & Sokal 1973).

Many of the data analyses were performed using the BIOSYS-1 program of Swofford & Selander (1981).

3 RESULTS

Banding patterns for all loci surveyed were interpreted as allelic variation at Mendelian loci (Harris & Hopkinson 1976). Laboratory crosses involving these and similar patterns in *Armadillidium vulgare* and other species of oniscids support this interpretation (Sassaman 1978a, 1978b, 1979; Sassaman & Garthwaite 1980, 1981, 1985; Garthwaite & Sassaman 1985). Alleles were expressed as deviation in mm from the most common North American allele which is designated 100. Allele frequency data and sample sizes for all 157 collections are summarized in Appendices 2A-F.

The relationships between allele frequency and latitude for the most common allele (100) at the GPI and PGM loci are shown in Fig. 2. Correlation coefficients for the relationship for GPI were –0.654 for Europe and –0.627 for North America, both of which are very highly significant ($P \ll 0.01$). For PGM allele 100, the coefficients were –0.198 for Europe, which is marginally significant (P is approximately equal to 0.05), and 0.161 for North America, which is non-significant ($P > 0.05$).

Regional averages of sample sizes and values for average number of alleles per locus and average observed heterozygosity are listed in Table 2. Both heterozygosity and mean number of alleles per locus are quite uniform across all regions with no consistent pattern to the variation present among regions.

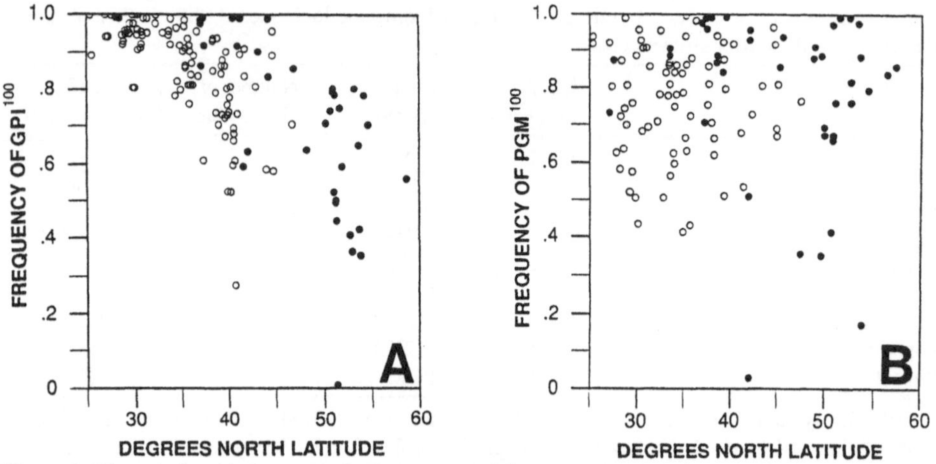

Figure 2. The relationship between the frequency of the most common allele (100) and latitude for A) GPI and B) PGM in population samples of *Armadillidium vulgare*. Closed circles = European populations, open circles = North American populations.

Table 2. Measures of genetic variability in population samples of *Armadillidium vulgare* averaged over regions of North America and Europe for both eight- and six-loci. Average standard errors are in parentheses.

Region		Number of populations	Mean sample size	Mean number of alleles	Mean observed heterozygosity
Eight loci examined:					
Europe:	north	13	63.5	1.4	.136
			(1.2)	(0.2)	(.071)
	south	14	60.7	1.8	.141
			(0.5)	(0.3)	(.064)
North America:	northeast	7	61.7	1.5	.142
			(0.3)	(0.2)	(.071)
	southeast	14	65.3	1.8	.171
			(1.1)	(0.2)	(.062)
	northwest	4	78.2	1.9	.169
			(2.8)	(0.2)	(0.62)
	southwest	3	65.9	2.1	.213
			(3.1)	(0.3)	(.063)
Six loci examined:					
Europe:	north	24	60.7	1.4	.127
			(0.4)	(0.2)	(.080)
	south	18	55.1	1.6	.107
			(0.4)	(0.3)	(.066)
North America:	northeast	20	70.2	1.5	.116
			(0.2)	(0.2)	(.066)
	southeast	27	70.3	1.8	.119
			(0.3)	(0.2)	(.063)
	northwest	22	69.9	1.7	.142
			(0.6)	(0.2)	(.066)
	southwest	32	79.1	1.8	.129
			(1.1)	(0.3)	(.057)

Of the 673 tests performed for conformance to Hardy-Weinberg equilibria, 48 were significant ($P < 0.05$) which is substantially more than the 33 expected significant deviations at this level due to chance departures. When the significance level is modified for multiple tests (Cooper 1968) six of the 673 tests show significant deviations in observed genotype frequencies from those expected under equilibrium conditions ($P < 0.05$). These deviations are at the IDH-1 locus in samples from Palo Alto, California; Guaymas, Mexico; and Toulouse, France, the PGM locus in samples from Praia de Mira, Portugal and Eugene, Oregon, and the GPI locus in a sample from the Chiricahua Mountains of Arizona. All six deviations were due to heterozygote deficits.

The inbreeding coefficient, F_{is}, is used here as an additional measure of departure from random mating within populations. In completely panmictic populations F_{is} is equal to zero. Positive deviations from zero indicate an excess of homozygotes, whereas negative deviations indicate a homozygote deficiency. F_{is} values for each polymorphic locus in each region, along with their X^2 tests of significance and means over loci are presented in Table 3.

Values of F_{is} for specific loci within regions in most cases show only slight departure from zero. In only two of 43 cases is the F_{is} coefficient significant ($P < 0.05$), these are at the MPI locus in northern Europe and at the AAT–1 locus in southern Europe. Using Cooper's (1968) correction for multiple tests none of the F_{is} coefficients for individual loci are significant. The departures of F_{is} from zero (Table 3) are about equally divided between positive and negative values and the means over loci, which probably have greater biological meaning than values for individual loci, are all non significant.

Table 3 also lists mean values of F_{st} for each of the six regions. Here, F_{st} is used as a measure of the relative amount of population (deme) differentiation within regions. In this study we use the criteria suggested by Hartl (1981) for the qualitative interpretation of the mean F_{st} values. Thus, the range from 0–0.05 indicates little genetic differentiation, 0.05–0.15 moderate differentiation, 0.15–0.25 great differentiation and > 0.25 very great differentiation. By these criteria, population differentiation in southeastern, northwestern and southwestern North America is moderate. Population differentiation in northeastern North America is great and in Northern and Southern Europe is very great.

From Table 3, the partitioning of genetic divergence among populations is greatest in southern Europe where 41% of the total genetic variability is between populations, leaving only 59% within populations ($1-F_{st}$). Northern Europe shows the same trend but with more variability within populations (74%) and less between. In North America about 85% or more of the total genetic variability is within populations and in the northwestern region only 7% of the total variation is between populations.

Wright's (1978) coefficients of allele frequency variance averaged over the six loci examined for most population samples were calculated for all pair-wise combinations of regions and are presented in Table 4.

There are some clear trends indicated by the data in Table 4. In general, genetic divergence is very much greater among populations within regions (F_{pr}) than it is between regions (F_{rt}). Furthermore, in these pair-wise comparisons, F_{pt} is not much greater than F_{pr}, again indicating that most of the variability is within regions. The greatest amount of inter-regional genetic divergence is seen between northern Europe

Table 3. The fixation index F_{is} with its X^2 for each locus averaged over populations of *Armadillidium vulgare* within regions of Europe and North America . Also given are the mean F_{is} and F_{st}.

Region		Locus	F_{is}	X^2	df	P	F_{st-}
Europe:	north	AAT–1	–.137	1.140	46	ns	
		GPI	.021	.027	69	ns	
		IDH–1	–.043	.113	69	ns	
		MDH–1	.057	.198	46	ns	
		MPI	.142	25.386	12	.013	
		PEPGL	.034	1.909	24	ns	
		PGM	.034	.071	24	ns	
		Mean	.071				.262
	south	AAT–1	.227	51.065	34	.030	
		GPI	–.072	5.137	68	ns	
		IDH–1	.230	52.424	68	ns	
		MDH–1	–.127	15.984	51	ns	
		MDH–2	–.005	.021	13	ns	
		MPI	.192	31.334	52	ns	
		PEPGL	–.024	.490	39	ns	
		PGM	–.043	1.832	34	ns	
		Mean	.016				.408
North America:	northeast	AAT–1	–.030	1.263	19	ns	
		GPI	.005	.035	19	ns	
		IDH–1	.039	2.134	38	ns	
		MDH–1	–.017	.405	19	ns	
		MPI	–.050	2.160	12	ns	
		PEPGL	–.011	.105	6	ns	
		PGM	.056	4.400	19	ns	
		Mean	.018				.159
	southeast	AAT–1	.024	1.093	52	ns	
		GPI	–.057	6.167	78	ns	
		IDH–1	–.023	1.004	78	ns	
		MDH–1	.048	4.373	26	ns	
		MPI	.009	.066	13	ns	
		PEPGL	–.024	.467	26	ns	
		PGM	.009	.154	26	ns	
		Mean	.002				.065
	northwest	AAT–1	–.012	.221	42	ns	
		GPI	.017	.444	21	ns	
		IDH–1	–.017	.444	84	ns	
		MDH–1	–.020	.615	21	ns	
		MPI	.041	.526	6	ns	
		PEPGL	–.036	.406	3	ns	
		PGM	.067	6.904	42	ns	
		Mean	.031				.071
	southwest	AAT–1	–.046	5.359	62	ns	
		GPI	–.035	3.103	93	ns	
		IDH–1	.050	6.331	93	ns	
		MDH–1	.019	.914	31	ns	
		MPI	–.014	.039	4	ns	
		PEPGL	.157	4.881	4	ns	
		PGM	–.011	.306	62	ns	
		Mean	.002				.138

Table 4. Wright's (1978) coefficient of allele frequency variance for populations within regions (F_{pr}), populations within the total (F_{pt}), and regions within the total (F_{rt}) for pair–wise comparisons among regions for American and European samples of *Armadillidium vulgare*. Coefficients are averaged over six loci (AAT–1, MDH–1, MDH–2, GPI, PGM, and IDH–1). N.A. = North America.

Regions compared		Number of populations	F_{pr}	F_{pt}	F_{rt}
North Europe vs:	south Europe	42	.330	.367	.056
	northeast N.A.	44	.235	.243	.011
	southeast N.A.	51	.163	.214	.061
	northwest N.A.	46	.176	.191	.018
	southwest N.A.	56	.164	.224	.071
South Europe vs:	northeast N.A.	38	.323	.334	.016
	southeast N.A.	45	.236	.248	.016
	northwest N.A.	40	.249	.257	.010
	southwest N.A.	50	.226	.225	.000
Northeast N.A. vs:	southeast N.A.	47	.124	.147	.026
	northwest N.A.	42	.140	.138	.000
	southwest N.A.	52	.131	.152	.024
Southeast N.A. vs:	northwest N.A.	49	.067	.077	.011
	southwest N.A.	59	.069	.080	.012
Northwest N.A. vs:	southwest N.A.	54	.081	.095	.015

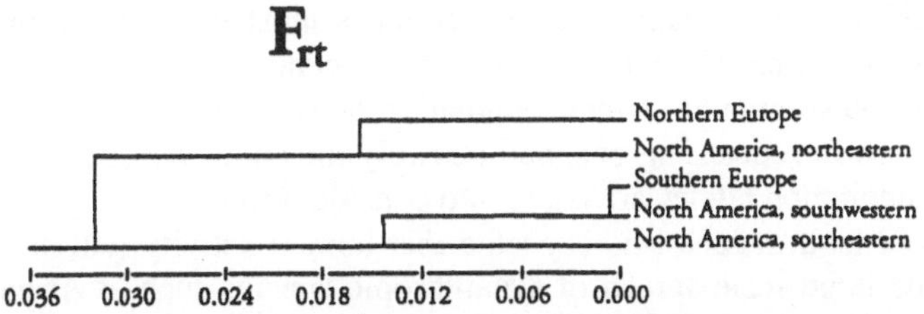

Figure 3. Cluster diagram based on the pair-wise values of Wright's (1978) coefficient of allele frequency variance for regions within the total (F_{rt}). The values (listed in Table 4) were clustered using the unweighted pair-group method, arithmetic averages (UPGMA).

and southern Europe and between northern Europe and southeast and southwest North America (6-7%). Divergence is low between northern Europe and northeastern North America, about 1%. Southern Europe and southwestern North America show no genetic divergence and that between southern Europe and the remaining North American regions ranges from one to about two per cent. Within North America divergence ranges from zero between the Northeast and the Northwest to about 3% between the Northeast and the Southeast. A cluster diagram summarizing the pairwise values of F_{rt} listed in Table 4 is shown in Figure 3.

4 DISCUSSION

One of the major differences in the patterning of genetic variation between North

American and European populations of *Armadillidium vulgare* observed in this study is the relative uniformity of genetic variation throughout North America in comparison to Europe, where allele frequencies are spatially much more variable (see Figs 4-8). The largest inter-regional genetic divergence is between northern and southern Europe. For many loci, large changes in allele frequencies occur over relatively short distances in Europe, while this is seldom, if ever, the case in North America. These results for *A. vulgare* are very similar to those obtained in a genetic analysis of European and New Zealand populations of chaffinches where genetic differentiation between northern and southern regions of the source continent (Europe), due to barriers affecting gene flow, was found to be much greater than that found in the descendent populations in New Zealand, which were known to be derived from a single European region (Baker 1992).

The absence of any significant departure of F_{is} in North American populations of *Armadillidium vulgare*, in conjunction with the relative uniformity of frequency distributions among these populations (Figs 4-8), suggests that random mating is complete within these populations and that there is a high level of gene flow among them. The small number of significant departures of F_{is} from zero in Europe and the six significant χ^2 values for departure of sample allele frequencies from Hardy-Weinberg expectations for both Europe and southwestern North America are all in the direction of homozygote excess in these samples. This finding suggests that the departures of genotype frequencies from expected values for these samples may be due to the inclusion of individuals from more than one non-interbreeding subpopulation in the samples (Wahlund 1928). This Wahlund effect is most likely to be observed when sampling is from areas showing the most heterogeneity among populations. In this study, the greatest amount of inter-population heterogeneity was found in the northern and southern European regions, and possibly the southwestern North American region when uncommon alleles are taken into consideration.

Within newly colonized land masses it is expected that human activity will have been a major factor in the large scale spread of synanthropic species such as *Armadillidium vulgare* from previously established populations. Local spread may occur naturally and mass movement for *A. vulgare* is well documented (Lokke 1966). Whatever the means of spread, in the case of adult gravid females of most of the larger oniscid isopods, for the reasons previously stated, a single founder is likely to carry with it a large proportion of the genetic variability present in the parent population (Sassaman 1978b; Johnson 1982), thus greatly reducing the impact of a population bottleneck on the genetic composition of derived populations. This is in contrast to many other organisms where founder effects in introduced populations are expected and do occur (e.g. Baker & Moeed 1987). The expected absence of the founder effect in *A. vulgare* is born out by the genetic data presented in this study. Average numbers of alleles and average heterozygosities are uniform throughout North America and are not substantially different from those found in Europe. Total number of alleles are also similar, and, in fact, two alleles have been found in North America which have not yet been encountered in Europe. Alleles which are thus far unique to Europe are concentrated in the eastern Mediterranean area.

Although *Armadillidium vulgare* thrives in anthropogenic environments, in the absence of competitors it is also able to colonize undisturbed areas. As there are few native terrestrial isopods in temperate North America, and no native species of *Ar-*

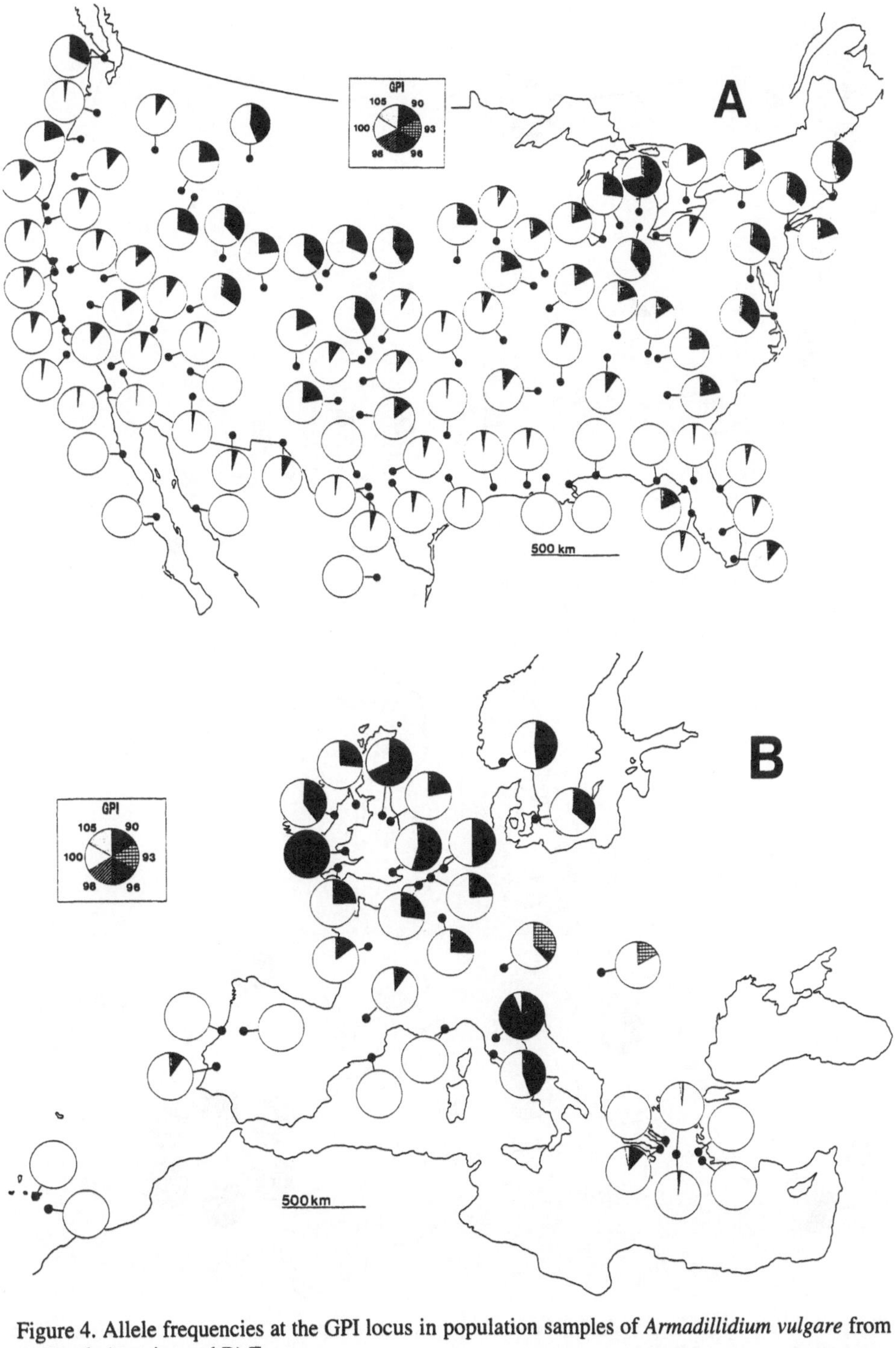

Figure 4. Allele frequencies at the GPI locus in population samples of *Armadillidium vulgare* from A) North America and B) Europe.

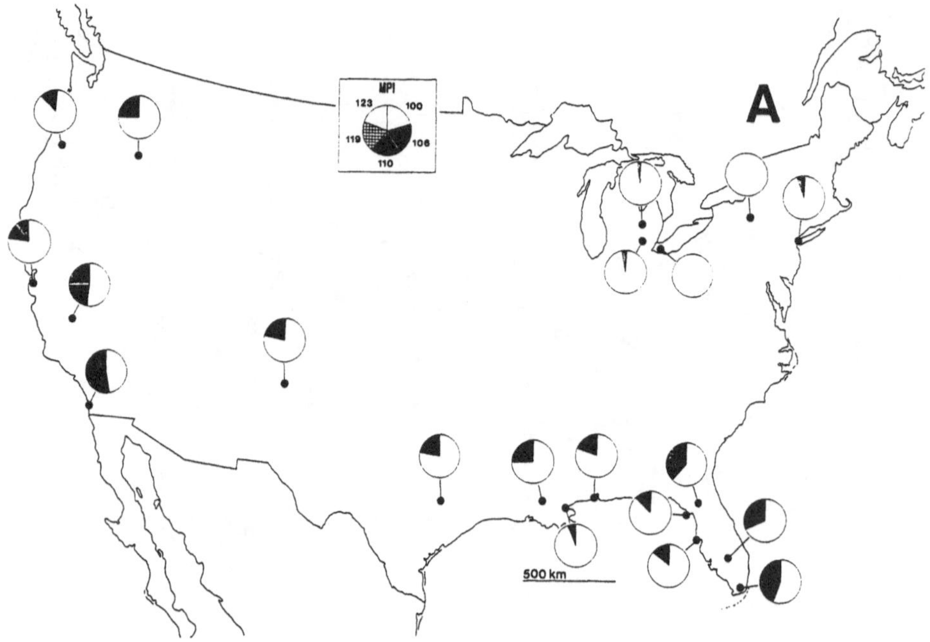

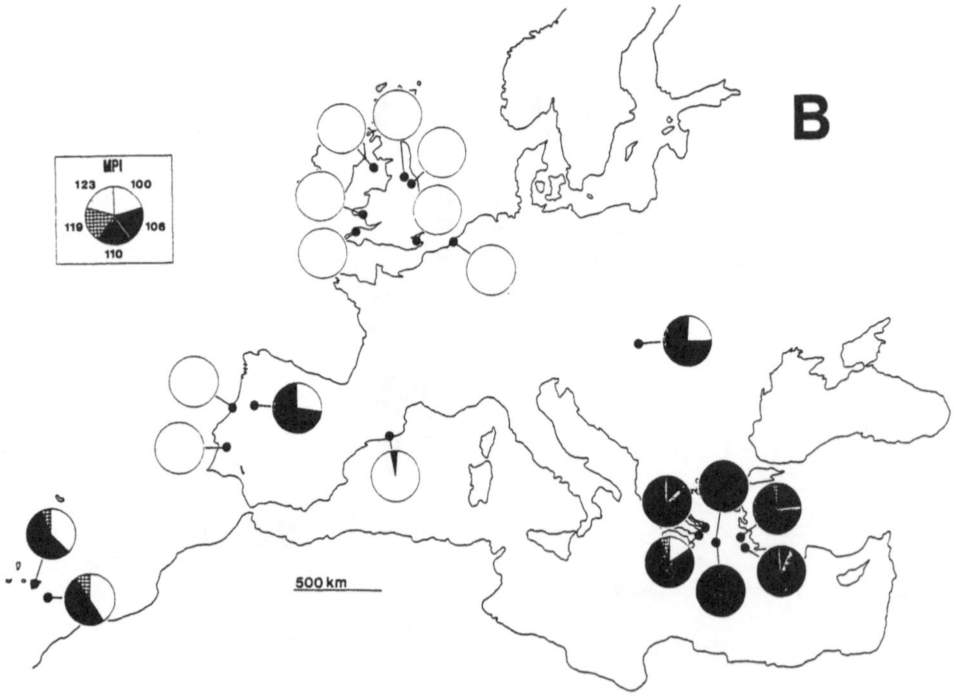

Figure 5. Allele frequencies at the MPI locus in population samples of *Armadillidium vulgare* from A) North America and B) Europe.

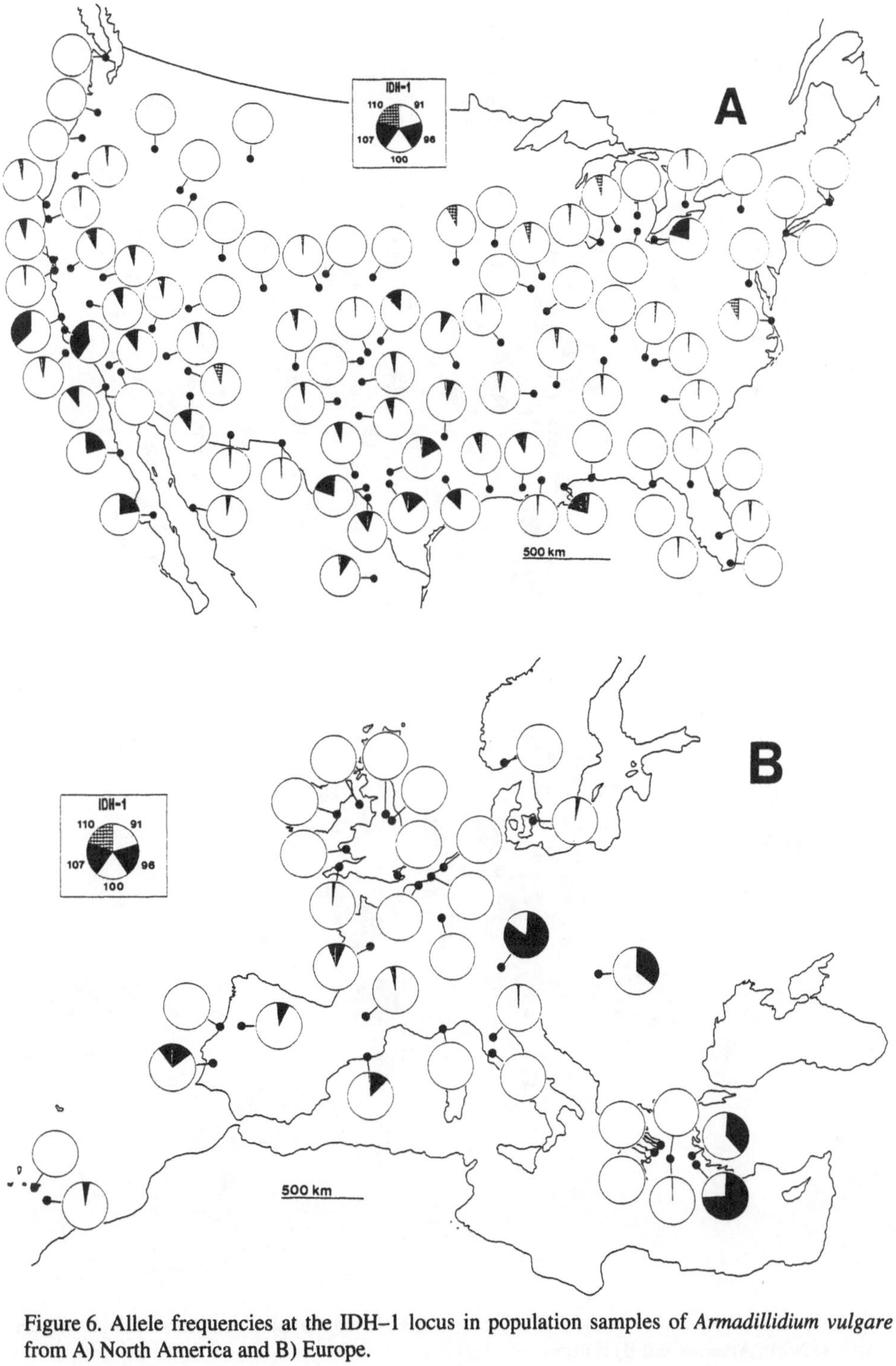

Figure 6. Allele frequencies at the IDH–1 locus in population samples of *Armadillidium vulgare* from A) North America and B) Europe.

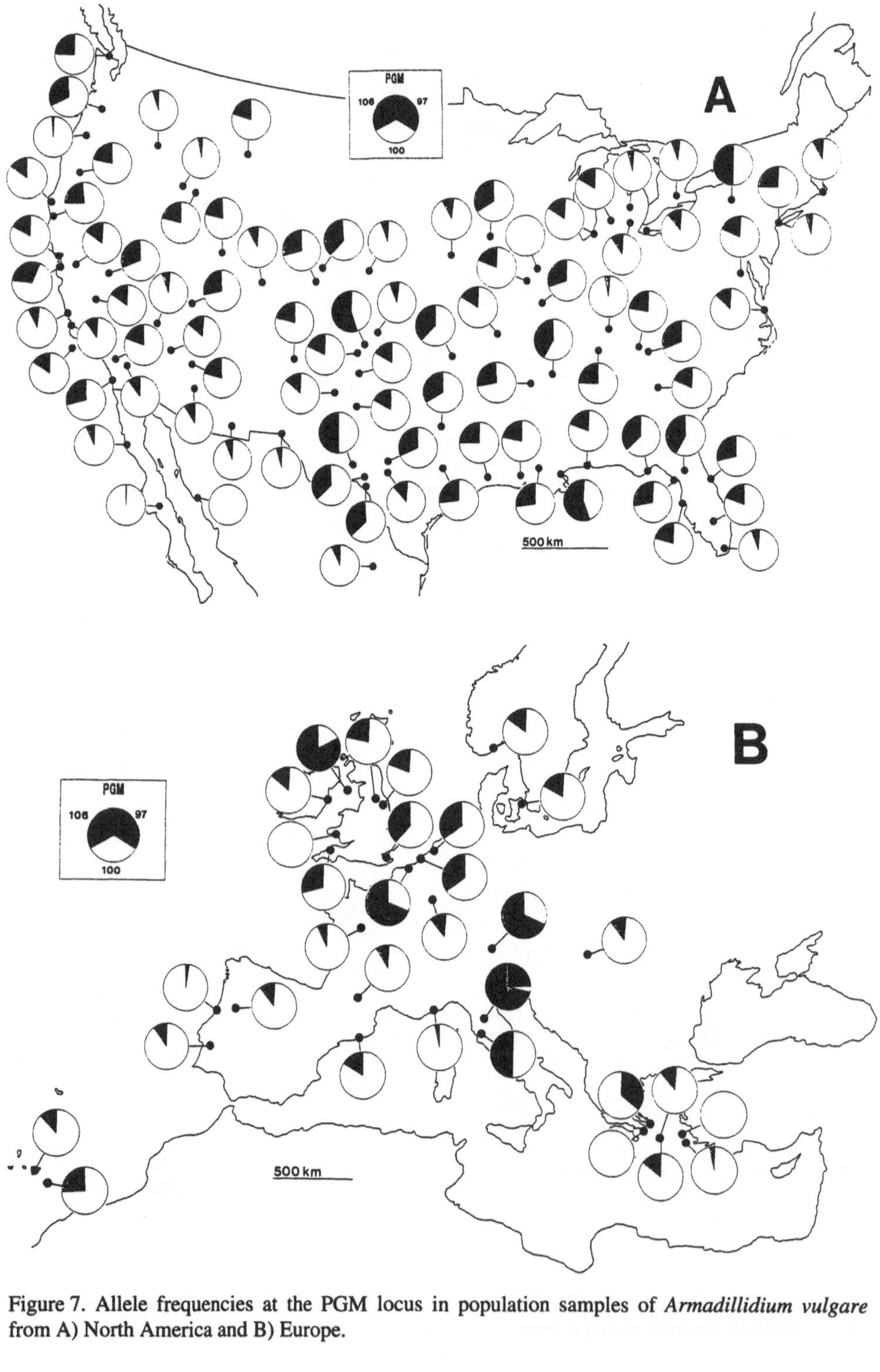

Figure 7. Allele frequencies at the PGM locus in population samples of *Armadillidium vulgare* from A) North America and B) Europe.

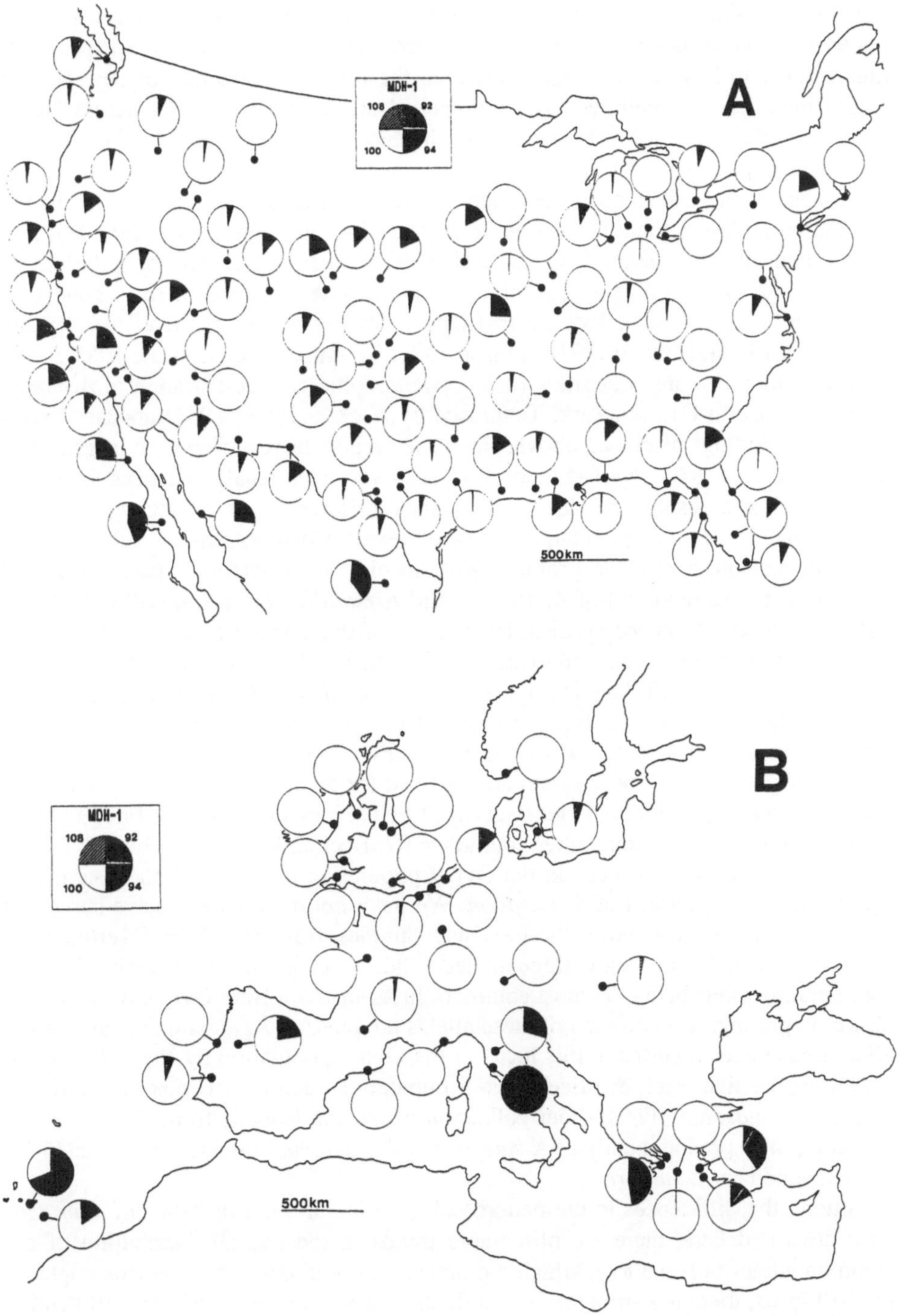

Figure 8. Allele frequencies at the MDH–1 locus in population samples of *Armadillidium vulgare* from A) North America and B) Europe.

madillidium (Van Name 1936), the advance of *A. vulgare* in North America would not have been impeded by competitors apart from other introduced oniscid species. If this resulted in low selection pressures then the initial colonization of new ground would have been uninhibited. The unimpeded spread of *A. vulgare* across North America from one or a few points of introduction would likely give the largely uniform genetic pattern observed in North America in this study. Conversely, the greater heterogeneity in gene frequencies among populations in Europe may be due to the fact that populations there have been established for longer periods of time and stochastic and selective factors have had a chance to act. However, once populations of oniscid isopods, including *A. vulgare*, become well established there is good evidence that alleles at some loci are resistant to frequency changes over many years (Sassaman 1978a; Howard 1981; Johnson 1985; Sassaman & Garthwaite 1985).

The results of a previous study of microgeographic genetic variation in *Armadillidium vulgare* in Overton Park, Tennessee by Beck & Price (1981) appear to contradict the findings discussed above, since they report large differences in gene frequencies at two loci (AAT–1 and AAT–2) over very small distances in North America. These results are due to frequency differences at a single population (Ovt–9) out of the nine populations they surveyed. However, their results at these two loci indicate a possible problem with species identification. Specifically, our surveys (data not reported) of *A. vulgare* and *Armadillidium nasatum* Budde-Lund, 1885, a second introduced species, from many of the same sites surveyed by Beck and Price indicate that Beck and Price probably included *A. nasatum* in their samples of *A. vulgare*. In North America, the AAT-1 locus in *A. vulgare* is dominated by a single allele (our allele 100, Beck & Price's 100) with only low frequencies of other alleles. Our samples of *A. vulgare* from Overton Park (collected at sites Ovt–2, Ovt–3, and Ovt–5 of Beck & Price) are all fixed for allele 100. *Armadillidium nasatum*, however, is, in all populations surveyed by us, including Beck & Price's Ovt–9 population, fixed for another allele (our allele 97, Beck & Price's 94). Beck & Price's allele frequencies at this locus in the Ovt–9 population indicate that their sample is predominantly comprised of *A. nasatum*. We have not done an extensive survey of AAT–2 in *A. vulgare*, never-the-less, our samples from the three Overton Park populations listed above show three alleles at this locus in this species with the fastest migrating allele being the most common. In *A. nasatum* from Ovt–9, only the two slower migrating, less common of these alleles are present. Again, on the basis of the allele frequencies reported at this locus in the Ovt–9 population by Beck & Price, it would appear that Beck & Price's Ovt–9 population sample is composed predominantly of *A. nasatum*. Finally, our collection of oniscid isopods from the Ovt–9 site was comprised predominantly of *A. nasatum* and *Cylisticus convexus* (De Geer 1778) and contained no *A. vulgare*.

Among the differences in the patterns of genetic variation in *Armadillidium vulgare* described here, there are observable trends in the overall distribution of the common alleles at some loci, which are present on both continents. For example, at the GPI locus, there is a strong north-south cline in the frequency of allele 100 with a decrease in frequency of this allele towards the north. This cline is apparent in both Europe and North America (Fig. 4). The magnitude of this cline can be measured by the correlation of the frequency of GPI allele 100 with latitude. For Europe, the Pearson product-moment correlation coefficient (r) is -0.654 ($P \ll 0.01$) and for North

America, r = –0.627 (P << 0.01) (Fig. 2). Though much less pronounced, similar clines occur in the distribution of less common alleles at other loci (MPI allele 110, Fig. 5; and IDH-1 allele 107, Fig. 6). In contrast to the GPI locus, common alleles at the PGM locus appear to be totally random in geographic distribution (Fig. 7), with little or no correlation between latitude and frequency of PGM allele 100 (r = –0.198 for Europe, P is approximately equal to 0.05; and r = 0.161 for North America, P > 0.05; Fig. 2). A north-south latitudinal cline has, however, been observed in alleles at the PGM locus in three other isopod species: *Asellus aquaticus* (L.), *Proasellus meridianus* Racovitz, and *Proasellus coxalis* Dollfus (Verspoor 1986). At the PEPGL locus in *A. vulgare*, in both Europe and North America, allele frequencies of the two common alleles are rather constant among most populations with both alleles present at fairly high frequencies.

These spatial patterns of allele frequency distribution can be explained by invoking non-historical factors such as selection and genetic drift. Verspoor (1986) has suggested temperature as the selective agent in producing the allele frequency cline at the PGM locus in the *Asellus* and *Proasellus* species. If selection is a factor in producing the latitude related clines seen at several loci in *Armadillidium vulgare* we are not at present able to identify which components of the environment may be the agent of selection. The random spatial pattern of allele frequency distribution at the PGM locus in relation to latitude is, perhaps, more suggestive of genetic drift as a determining factor. If selection is involved, then the spatial scaling is quite different to that seen for the GPI locus. In the case of the distribution of the frequencies of the two common PEPGL alleles, in general, these seem to be held within certain limits within many populations in both Europe and North America: approximately 60% for allele 100 and 40% for allele 109. Because of this, it is tempting to invoke balancing selection as the cause of this stability. However, we have no direct evidence that such is the case and have no knowledge of what such a selective agent might be.

Selection has been implicated in the maintenance and patterning of genetic variation at specific loci in several isopod species. Howard (1981) found long term stability in frequencies of alleles governing body color in *Armadillidium vulgare*. Presumably allele frequencies at this locus are maintained in stasis by gene flow and/or balancing selection. Haldane (1962) showed evidence that indicated that selection operating on young individuals was the stabilizing factor for this polymorphism. On what is probably the homologous locus in *Armadillidium nasatum*, maintenance of the polymorphism has also been attributed to selection (Adamkewicz 1969). In the related *Venezillo parvus* (Budde-Lund 1885) (= *V. evergladensis* Schultz 1963), stable allele frequencies at this same or a similar locus were, by experimental evidence, also shown to be maintained by selection operating on young individuals (Johnson 1985).

Although there is some evidence that allozyme frequencies in terrestrial isopod populations are stable at least over short periods of time (Sassaman 1978; Sassaman & Garthwaite 1985), the extent to which allozymes are influenced by selection in species in general has been and remains a debated question. Enhanced fitness of individuals possessing certain alleles as homozygotes or in heterozygous combination have been demonstrated in other organisms (Powers et al. 1979; Koehn et al. 1980; Watt et al. 1983; see Endler 1986 for further examples). Within the Isopoda, studies by Heath et al. (1988) have shown that a di-allelic polymorphism at the GPI locus in

Sphaeroma rugicauda (Leach) was maintained by differential transmission of male alleles to their offspring which was dependent on the genotype of the maternal female. In *Asellus aquaticus* low temperatures have been implicated in selection against heterozygotes at an amylase locus (Christensen 1977). Other studies of *A. aquaticus* have shown higher fecundity to be associated with certain GPI genotypes in females (Shihab & Heath 1987). However, such a finding could not be demonstrated for this same polymorphic locus in *Sphaeroma rugicauda* (Heath et al. 1988). In all of these examples of selection affecting allozymes, selection may alter frequencies from one generation to the next, however, it is expected that the long term effects will be to maintain the polymorphism. The majority of studies of most organisms, however, have failed to demonstrate departure from neutrality for allozymes (Singh & Long 1992) and this is consistent with model predictions which assume only random drift and mutation to be affecting allozyme variation (Kimura 1983 Nei 1983).

A locus by locus comparison of the genetic structure of *Armadillidium vulgare* in the six regions included in this study shows numerous similarities between northern Europe and northeastern North America and southern Europe and southwestern and southeastern North America. At the AAT–1 locus, allele 104 was found only in southern Europe, and in North America was commonly encountered along the west coast and to a more limited extent in the southeast (Fig. 9). At the GPI locus, allele 96 was predominantly found in the north in Europe while the alternate common allele, 100, was predominantly southern in its distribution (Fig. 4). In North America, GPI allele 96 was found to be more common in the north and northeast while allele 100 dominates in the southwest and southeast (Fig. 4). Additionally, GPI allele 90, which was found only in southern Europe, was found only in California in North America (Fig. 10). For the IDH–1 locus, alleles 107 and 96 are largely southern in their distribution in Europe and are largely southeastern and southwestern in their distribution in North America (Fig. 6). At the MDH–1 locus, allele 92 is more commonly encountered in the south in Europe and the west in North America (Fig. 8). At the MPI locus, allele 106 is restricted to the south in Europe and was found only in California in North America (Fig. 5). Similarly, at the PGM locus, allele 97 is restricted entirely to the south in Europe and was found only in the west in North America (Fig. 11).

Wright's F-statistics for pair-wise comparisons of regions summarize and reinforce these individual locus results for *Armadillidium vulgare*. These F-statistics show a distinct genetic similarity between northern Europe and northeastern North America and southern Europe and southwestern North America (Table 4, Fig. 3). Because the uncommon alleles discussed above have only slight effects on Wright's (1978) hierarchical F-statistics they serve as independent indicators of population origins for introduced species.

Human settlement of North America by Europeans has been one of two separate colonizations. Northern Europeans colonized northeastern North America, while the initial colonization of the southwest and southeast was by southern Europeans, primarily the Spanish. *Armadillidium vulgare* has, without doubt, been introduced into North America in conjunction with human migration from Europe. However, published collection records for *A. vulgare* in North America suggest a single focus of introduction, this being on the northeastern coast of the United States. Figure 12

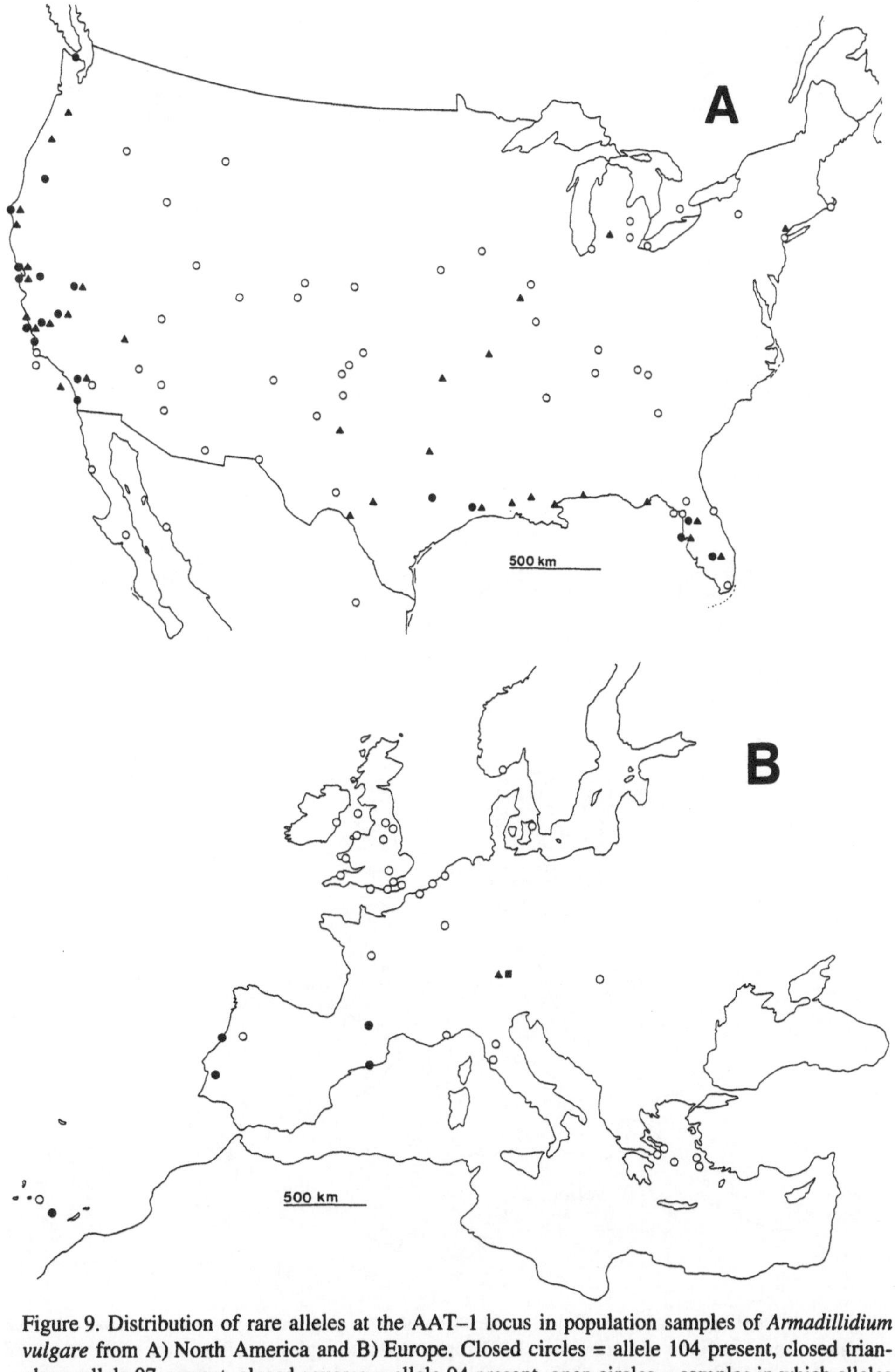

Figure 9. Distribution of rare alleles at the AAT–1 locus in population samples of *Armadillidium vulgare* from A) North America and B) Europe. Closed circles = allele 104 present, closed triangles = allele 97 present, closed squares = allele 94 present, open circles = samples in which alleles 104, 97, and 94 were not found.

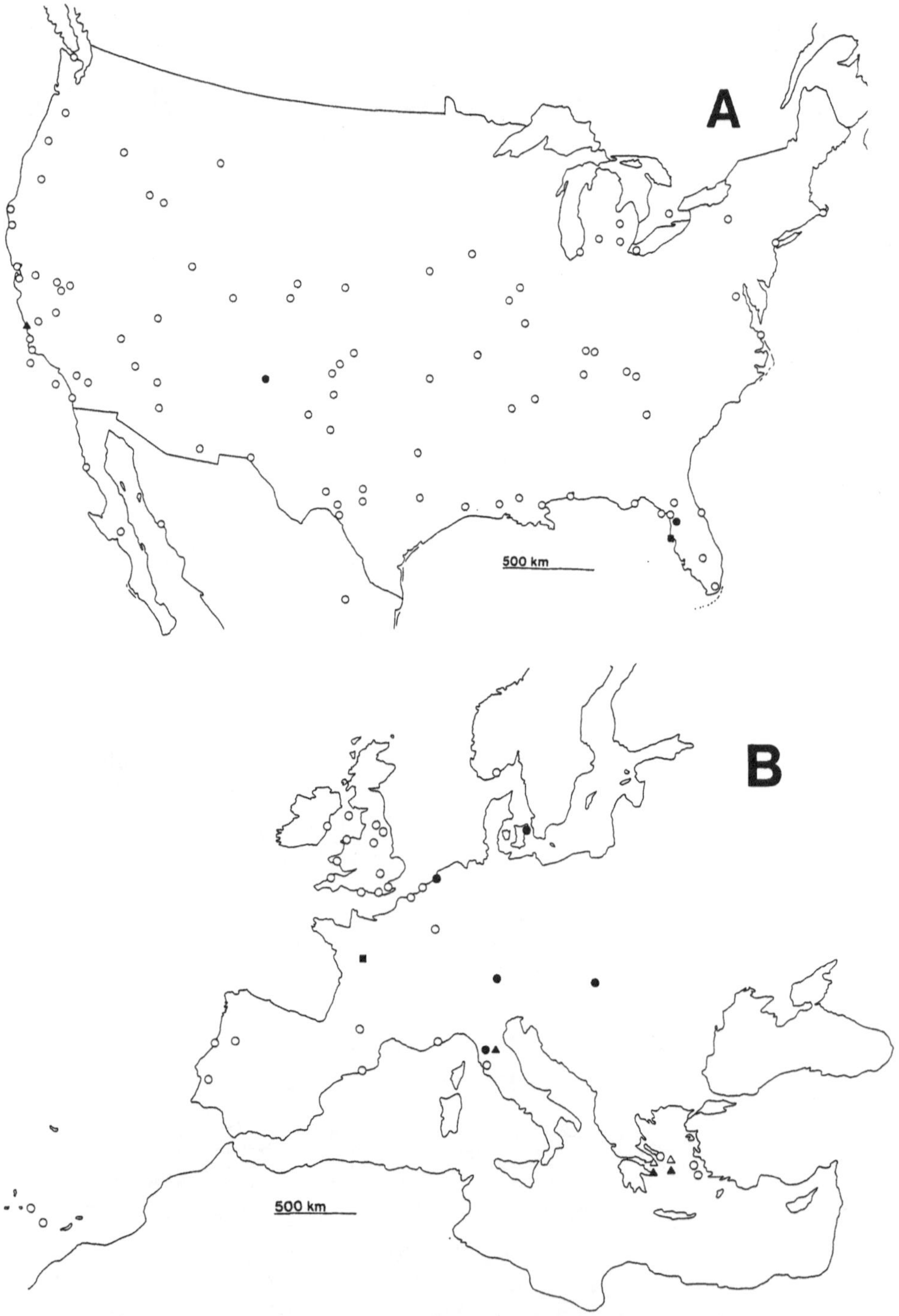

Figure 10. Distribution of rare alleles at the GPI locus in population samples of *Armadillidium vulgare* from A) North America and B) Europe. Open triangles = allele 105 present, closed squares = allele 98 present, closed circles = allele 93 present, closed triangles = allele 90 present, open circles = samples in which alleles 105, 98, 93, and 90 were not found.

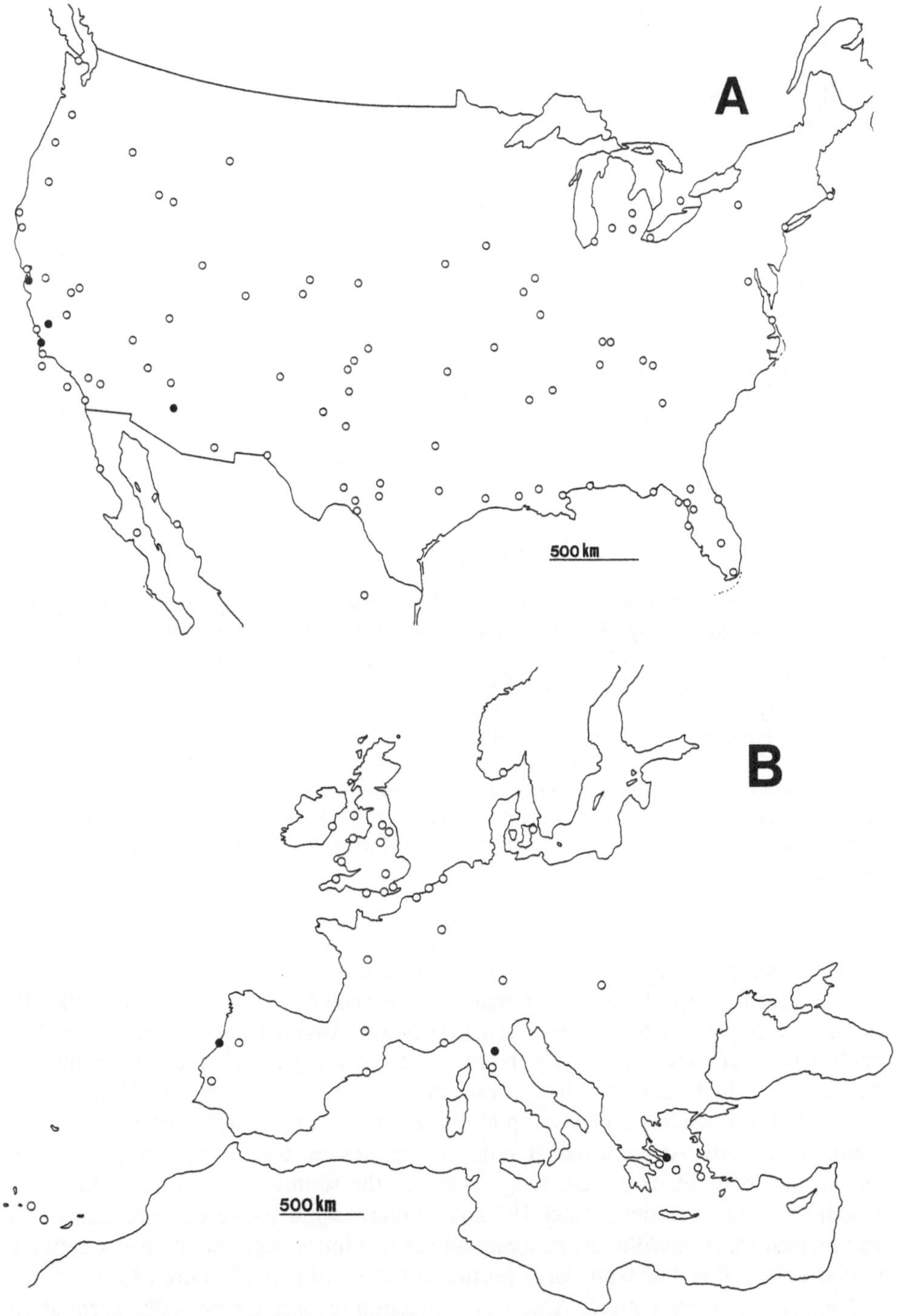

Figure 11. Distribution of allele 97 at the PGM locus in population samples of *Armadillidium vulgare* from A) North America and B) Europe. Closed circles = samples in which allele 97 was present, open circles = samples in which allele 97 was not found.

Figure 12. The introduction and spread of *Armadillidium vulgare* in North America based on published accounts (Say 1818; Gould 1841; DeKay 1844; Fitch 1856; Underwood 1886; Dollfus 1897; Paulmier 1905; Rathbun 1905; Richardson 1905; Pierce 1907; Cockerell 1912; Fowler 1912; Kunkel 1918; Longnecker 1923; Essig 1926; Walker 1927; Swenk 1929; Blake 1931; Arcangeli 1932; Brimley 1938; Miller 1938; Hatch 1939; Cockerell 1940; Mulaik & Mulaik 1942, 1943; Hatch 1947; Hatchett 1947; Eberly 1953; Muchmore 1957; Drummond 1965; Schultz 1965; Warburg 1965; Merriam 1971; Schultz 1975; Lane 1977; Sorensen & Burkett 1977; Schultz 1984). Solid symbols = collections reported prior to 1905, open symbol = collections reported after 1905. Closed triangles = collections reported from 1800 to 1850, closed circles = collections reported from 1851 to 1905, open triangles = collections reported from 1906 to 1950, open circles = collections reported from 1951 to present.

shows the temporal spread of *A. vulgare* throughout North America, based on published collection records, in approximately 50 year intervals starting from 1800. By 1850, *A. vulgare* had been introduced into North America but was restricted to a small Section of New England. By the turn of the century it had spread over much of the eastern United States and had advanced as far west as Ft. Worth, Texas. However, not until relatively recently (up to 1950) has *A. vulgare* achieved the wide distribution in North America that it presently possesses, and it appears to have expanded into parts of its present range (such as the southwestern United States and Florida) even more recently, since 1950. A similarly rapid and recent spread has been documented for *Armadillidium nasatum* which was introduced into North America at around 1900 and now inhabits large portions of the continent (Schultz 1961).

The paucity of early North American collection records for oniscids, particularly for Mexico, does not allow us to rule out additional foci of introduction. However, the fact that other species of European oniscids (e.g. *Porcellio laevis* Latreille 1804; *Porcellio scaber* Latreille 1804; *Porcellionides* spp. and *Cylisticus convexus*) were being collected from large areas of western North America long before *Armadillid-*

ium vulgare was appearing in these areas (Richardson 1905) suggests that the pattern of introduction of *A. vulgare* summarized in Figure 12 is real.

The genetic results presented in this study support an historic component to the patterning of genetic variation in *Armadillidium vulgare* in North America and argue for two separate introductions. Northeastern North American populations may be similar to north European populations because *A. vulgare* was introduced into North America by northern Europeans, whereas southwestern North American populations may be similar to south European populations because *A. vulgare* was introduced into the southwest by southern Europeans. However, since the climate and other environmental characteristics of northeastern North America are similar to Northern Europe and those of southwestern North America are similar to southern Europe, the observed similarity of allele frequencies in the northeast and northern Europe and the southwest and southern Europe is expected under both historic and adaptive theories.

The similarities found in this study between *Armadillidium vulgare* allele frequencies in northern Europe and the northeast and southern Europe and the southwest are not exact. If allele frequencies are primarily the result of selective pressures, then changes in their patterning would be expected in areas of introduction. However, differences in the patterning of allele frequencies between source and derived populations could occur under an entirely historic explanation for a number of reasons: 1) Even if selective pressures are relaxed in areas of colonization, leaving only stochastic processes to affect allele frequencies, the allele frequencies of source populations may not necessarily remain static but may change over time due to selective or non-selective forces. In this case, derived populations may more closely match source populations at the time of colonization than present day populations at the source area (Todd 1977). 2) Due to founder effect, derived populations may not match source populations even at the time of introduction, although, as discussed above, this is less likely for oniscid isopods than for most other organisms. 3) Because of the vagaries of human migration, lack of detailed information on human migration patterns can cause the genetic composition of synanthropic species to appear falsely anomalous (Todd et al. 1976; Morrill & Todd 1978). 4) The influence of human migration on the distribution of genes in synanthropic organisms need not be uniform across all genes. The known pattern of human migration can result in vastly different patterns of genetic variation at different loci in synanthropic species (Todd 1977). 5) Human movement does not always guarantee the movement or establishment of the synanthropic organism (N.B. Todd & L.M. Todd 1976). 6) Finally, the exact source of introduced species is seldom known and introduced individuals may represent a mixture of individuals from several or many source populations. Thus the data presented here do not allow us to identify specific source localities in Europe on the basis of the presence of alleles and their frequencies and such an attempt would most likely be futile.

The genetic structure of other synanthropic species which have been introduced into new areas by the actions of man have been shown to have strong historical components (e.g. Todd 1977; Morrill & Todd 1978; Bryant et al. 1981; Enckell et al. 1986; Baker 1992; Singh & Long 1992). Due to the close relationship between domestic cats and humans, work on cat coat genes provides an example of a system which, although under the influence of strong selection in certain ways (Todd 1977), is probably as close to being predominantly influenced by strictly historical factors as will be found. Studies of domestic cat coat genes throughout Europe and North America have demonstrated the existence of a pattern to these genes in North America which correlates well with the patterns of human colonization and migration throughout the continent (Todd et al. 1976; Morrill & Todd 1978). Overall, the distribution of cat coat genes in North America suggests a dual introduction of domestic cats into the continent. Southwestern and Mexican cats have a genetic makeup similar to Spanish cats due to their introduction by Spanish settlers, whereas the cats of northeastern North America are similar to English cats, also presumably by descent (Todd et al. 1976; Morrill & Todd 1978). The gene pools of the two groups of cats are apparently now mixing in parts of Texas and California (Todd et al. 1976; Morrill & Todd 1978). The extent to which the spatial patterning of cat coat genes is influenced by human migration in North America can often be seen on a very detailed scale (e.g. see Halpine & Kerr 1986).

A close correlation between human migration patterns and the worldwide pattern of genetic variation in *Drosophila melanogaster* has also been observed (Singh & Long 1992). In this case, the reestablishment of allozyme and other clines in areas of relatively recent introduction indicates that evolutionary change in response to selective factors can be, at least in some organisms, very rapid (Singh & Long 1992). However, in at least one synanthropic organism, allozyme frequency differences among derived populations may have persisted for hundreds of years (Enckell 1986).

This study has demonstrated that the level of genetic variation in *Armadillidium vulgare*, both at the population and regional level, is sufficient for studies on the adaptive significance of this variation. Since *A. vulgare* has been distributed throughout the world by Europeans, this system allows a number of independent tests of the selective importance of genetic variation in this species. Due to similarities in climate between likely source areas in Europe and areas of colonization in North America, the cause of the genetic similarity between these regions is, unfortunately, ambiguous. However, if there has, in fact, been a single focus of introduction in *A. vulgare* in North America as indicated in Figure 12, then the involvement of selective forces in the patterning of genetic variation in this species would be likely.

In the future, we plan to concentrate on analyzing populations of *Armadillidium vulgare* from a number of regions where the source of human colonization (and thus isopod colonization) has a climate significantly different from the colonized area (for example, southern South America) as these situations will provide the most direct test of the selective value of genetic variation in this species.

ACKNOWLEDGEMENTS

We thank the following persons for their assistance in collecting isopod samples included in this study: J.S. Applegarth, A. Brito, T. Brown, C. Burmester, P.G. Frank, E. Garthwaite, B. Hanks, P. Harding, C. Howes, P. Juchault, R.B. King, V. Lance, B.L. Lawson, S.M. Lawson, D.L. Martin, G. Merker, P. Moler, N. Moncrief, I.K. Morgan, Y. Murata, G.T. Owen, F. Parker, D. Richardson, S. Sfenthourakis, K. Szlavecz, S. Taiti, D. Terrone, J.L. van Tassell.

APPENDIX 1

Collection data for population samples of *Armadillidium vulgare*.

Code	Geographic division: First	Second	Locality
Austria:			
INN	Tirol		Innsbruck
Belgium:			
ARL	Luxembourg		Arlon
BEL	West–Vlaanderen		Nieuwpoort
Canada:			
BCA	Ontario	Brant	Brantford
MID	Ontario	Essex	Middle Island
Denmark:			
CDK	Zealand	København	Copenhagen
France:			
CFR	Poitou	Deux–Sèvres	Celles–sur–Belle
TFR	Gascogne	Haute–Garonne	Toulouse
Greece:			
ALO	Thessaly		Patitiri, Alonnisos Island
ATT	Central Greece		Attiki
EVI	Central Greece		Almiropotamos, Évvoia Island
GRC	Aegean Islands		Astipalaia Island
KOM	Aegean Islands		Komi, Tinos Island
LER	Aegean Islands		Gourna, Leros Island
PAT	Aegean Islands		Alykes, Pátmos Island
TIN	Aegean Islands		Kolymbithres, Tinos Island
Hungary:			
HUN	Pest		Budapest
Ireland:			
MIR	Dublin		Dublin
Italy:			
COR	Toscana	Firenze	Corbignano
PNM	Toscana	Grosseto	Grosseto
VEN	Liguria	Imperia	Ventimiglia
Mexico:			
CBC	Baja Calif. Norte		Catavina
GMX	Sonora		Guaymas
MTM	Nuevo Leon		Monterrey
PBC	Baja Calif. Sur		San Ignacio
Netherlands:			
ANL	Noord Holland		Amsterdam
HOL	Zeeland	Walcheren	Zoutelande
Norway:			
HNW	Aust–agder		Humleholmen
Portugal:			
ADS	Baixo Alentejo		Alcaser do Sal
PDM	Beira Litoral		Praia de Mira

Code	First	Second	Locality
Spain:			
FDO	Salamanca		Fuentes de Onoro
HOS	Barcelona		l'Hospitalet
JCA	Canary Islands		Grand Canary Island
TEN	Canary Islands		Tenerife Island
United Kingdom:			
BAR	North Lincoln		Barton on Humber
COG	Glamorgan		Cliffs of Gower
DON	South–West York		Doncaster
EEG	Cambridge		Elsworth
ETW	North Devon		Bideford
IOM	Isle of Man		Castletown
IOW	Isle of Wight		Bouldnor
IWA	East Kent		Iwade
KHB	East Kent		Herne Bay
KNA	Mid–West York		Knaresborough
LLO	Caernarvon		Llandudno
MFU	Glamorgan		Morfa–uchaf
RAV	North Lincoln		Ravendale
USX	East Sussex		Brighton
United States of America:			
ADI	AL	Mobile	Dauphin Island
CH	AZ	Cochise	Chiricahua Mts.
KAR	AZ	Mohave	Kingman
PRE	AZ	Yavapai	Prescott
SFC	AZ	Cochise	Chiricahua Mts.
TZ1	AZ	Maricopa	Tempe
TZ4	AZ	Maricopa	Tempe
ARK	AR		Northeast Arkansas
CAT	CA	Los Angeles	Avalon
CGC	CA	San Bernadino	Cushenbury Grade
CME	CA	Humboldt	Petrolia
DEM	CA	San Diego	Del Mar
GAV	CA	Santa Barbara	Gaviota
H41	CA	Madera	Oakhurst
IRS	CA	Mendocino	Laytonville
LFS	CA	Monterey	Lockwood
LKB	CA	Monterey	Lucia
NIC	CA	San Luis Obispo	Pismo Beach
ORO	CA	Tulare	Orosi
PAA	CA	Santa Clara	Palo Alto
PCO	CA	Riverside	Palm Canyon Oasis
SAR	CA	Riverside	Rubidoux
SCA	CA	San Joaquin	Stockton
SNR	CA	Tuolumne	Sonora
SRA	CA	Marin	San Rafael
SRI	CA	Santa Barbara	Santa Rosa Island
SSB	CA	San Luis Obispo	San Simeon
TIO	CA	Mono	Lee Vining
VZ	CA	Riverside	Riverside
FRU	CO	Mesa	Fruita
GMC	CO	Adams	Denver

Code	First	Second	Locality
KCO	CO	Weld	Kersey
FL1	FL	Citrus	Floral City
FL2	FL	Alachua	Gainesville
FL3	FL	Pinellas	Toytown dump
FL4	FL	Glades	Lake Okeechobee
FL5	FL	Citrus	Floral City
FL6	FL	Citrus	Yankeetown
FL7	FL	Levy	Cedar Key
FL8	FL	Pinellas	Dunedin
FL9	FL	Dade	Florida City
GAI	FL	Alachua	Gainesville
GFL	FL	Alachua	Gainesville
LAN	FL	Franklin	Lanark Beach
NSB	FL	Volusia	New Smyrna Beach
AGA	GA	Clarke	Athens
GFI	ID	Elmore	Glenns Ferry
JID	ID	Jerome	Jerome
SHI	ID	Gooding	Shoshone
MIL	IL	McDonough	Macomb
AIA	IA	Story	Ames
KKN	KS	Seward	Kismet
EHC	KY	Hart	Horse Cave
FOA	KY	Barren	Cave City
KKM	KY	Barren	Cave City
SBF	KY	Barren	Cave City
STR	KY	Edmonson	Huff
ENO	LA	Orleans	East New Orleans
LLA	LA	Lafayette	Lafayette
LSU	LA	East Baton Rouge	Baton Rouge
WH	MA	Barnstable	Woods Hole
BCM	MI	Calhoun	Battle Creek
NBM	MI	Berrien	New Buffalo
UMC	MI	Washtenaw	Ann Arbor
WSU	MI	Wayne	Detroit
SMS	MS	Franklin	Meadville
MTH	MO	Marion	Hannibal
SCF	MO	St. Louis	St. Louis
SLM	MO	St. Louis	St. Louis
SMO	MO	Greene	Springfield
CHS	MT	Park	Chico Hot Springs
INB	NB	Chase	Enders Reservoir
OMA	NB	Douglas	Omaha
LVN	NV	Clark	Las Vegas
PON	NM	Roosevelt	Portales
SFE	NM	Santa Fe	Santa Fe
BNY	NY		Bayside
CPW	NY	New York	Manhattan
INY	NY	Tompkins	Ithaca
RSI	OH	Ottawa	Rattlesnake Island
WSI	OH	Ottawa	West Sister Island
GOK	OK	Texas	Goodwell
GWO	OK	Texas	Guymon
TOK	OK	Tulsa	Tulsa

Code	First	Second	Locality
AOR	OR	Jackson	Ashland
GLD	OR	Clackamas	Gladstone
JSA	OR	Lane	Eugene
OR1	OR	Baker	Baker
CHW	TN	Davidson	Nashville
GTN	TN	Sevier	Gatlinburg
ORT	TN	Anderson	Oak Ridge
OVT	TN	Shelby	Memphis
AMA	TX	Potter	Amarillo
CTX	TX	Val Verde	Comstock
DFW	TX	Tarrant	Dallas–Ft. Worth Airport
DRT	TX	Val Verde	Devils River
JTX	TX	Kimble	Junction
LTX	TX	Lubbock	Lubbock
MEN	TX	Menard	Menard
PAS	TX	El Paso	El Paso
PRT	TX	Pecos	Sheffield
STX	TX	Sherman	Stratford
TEX	TX	Brazos	College Station
WTX	TX	Jefferson	Beaumont
MVW	UT	Piute	Marysvale
SQU	UT	Utah	Santaquin
ALX	VA	Fairfax	Alexandria
VB2	VA		Virginia Beach
SQW	WA	Clallam	Sequim

APPENDIX 2A

Allele frequencies for collections of *Armadillidium vulgare* from northern Europe. Refer to Appendix 1 for explanation of population codes. Sample sizes, in number of individuals, are in parentheses.

Locus/ Allele	Country/population code: NL ANL	BELG ARL	UK BAR	BELG BEL	DEN CDK	FR CFR	UK COG	UK DON
AAT–1	(104)	(74)	(149)	(24)	(45)	(52)	(67)	(117)
104	–	–	–	–	–	–	–	–
100	1.000	1.000	1.000	1.000	1.000	1.000	1.000	1.000
97	–	–	–	–	–	–	–	–
94	–	–	–	–	–	–	–	–
GPI	(104)	(74)	(149)	(24)	(45)	(52)	(67)	(117)
105	–	–	–	–	–	–	–	–
100	.500	.743	.775	.729	.644	.846	–	.564
98	–	–	–	–	–	.010	–	–
96	.495	.257	.225	.271	.344	.144	1.000	.436
93	.005	–	–	–	.011	–	–	–
90	–	–	–	–	–	–	–	–
IDH–1	(104)	(74)	(149)	(24)	(45)	(52)	(67)	(117)
110	–	–	–	–	–	–	–	–
107	–	–	–	–	–	.058	–	–

Locus/ Allele	NL ANL	BELG ARL	UK BAR	BELG BEL	DEN CDK	FR CFR	UK COG	UK DON
100	1.000	1.000	1.000	1.000	.967	.875	.918	1.000
96	–	–	–	–	.033	.067	.082	–
91	–	–	–	–	–	–	–	–
IDH–2	(104)	(74)		(24)	(45)	(52)	(67)	
A	.019	–		–	–	–	–	
102	–	–		–	–	–	–	
100	.981	1.000		1.000	1.000	1.000	1.000	
MDH–1	(72)	(74)	(149)	(24)	(45)	(52)	(67)	(117)
108	–	–	–	–	–	–	–	–
100	.868	1.000	1.000	.979	.933	1.000	1.000	1.000
94	–	–	–	–	–	–	–	–
92	.132	–	–	.021	.067	–	–	–
MPI			(149)					(117)
123			–					–
119			–					–
110			–					–
106			–					–
100			1.000					1.000
PEPGL		(74)	(136)	(18)		(40)	(60)	(101)
113		–	–	–		–	–	–
109		.297	.563	.639		.325	.317	.723
100		.689	.438	.361		.675	.683	.277
95		.014	–	–		–	–	–
PGM	(104)	(74)	(149)	(24)	(45)	(52)	(67)	(112)
106	.346	.122	.201	.688	.178	.077	.597	–
100	.654	.878	.799	.313	.822	.923	.403	1.000

Locus/ Allele	UK EEG	UK ETW	NOR HNW	NL HOL	HUN HUN	AUS INN	UK IOM	UK IOW
AAT–1	(39)	(46)	(68)	(23)	(17)	(50)	(37)	(66)
104	–	–	–	–	–	–	–	–
100	1.000	1.000	1.000	1.000	1.000	.870	1.000	1.000
97	–	–	–	–	–	.120	–	–
94	–	–	–	–	–	.010	–	–
GPI	(39)	(46)	(68)	(23)	(17)	(50)	(37)	(66)
105	–	–	–	–	–	–	–	–
100	.308	.750	.529	.761	.824	.620	.730	.402
98	–	–	–	–	–	–	–	–
96	.692	.250	.471	.239	–	.070	.270	.598
93	–	–	–	–	.176	.310	–	–
90	–	–	–	–	–	–	–	–
IDH–1	(39)	(46)	(68)	(23)	(17)	(50)	(37)	(66)
110	–	–	–	–	–	–	–	–
107	–	–	–	–	–	–	–	–
100	1.000	.978	1.000	1.000	.647	.170	1.000	1.000
96	–	.022	–	–	.353	.830	–	–
91	–	–	–	–	–	–	–	–

Locus/ Allele	UK EEG	UK ETW	NOR HNW	NL HOL	HUN HUN	AUS INN	UK IOM	UK IOW
IDH–2	(39)		(68)			(50)		
A	–		–			–		
102	–		–			–		
100	1.000		1.000			1.000		
MDH–1	(39)	(46)	(68)	(23)	(17)	(50)	(37)	(66)
108	–	–	–	–	.029	–	–	–
100	1.000	1.000	1.000	1.000	.971	1.000	1.000	1.000
94	–	–	–	–	–	–	–	–
92	–	–	–	–	–	–	–	–
MPI		(46)		(23)	(17)		(37)	(66)
123		–		–	–		–	–
119		–		–	–		–	–
110		–		–	.765		–	.212
106		–		–	–		–	–
100		1.000		1.000	.235		1.000	.788
PEPGL		(46)		(23)	(17)	(40)	(37)	(66)
113		–		–	–	–	–	–
109		.054		.370	.971	.900	.311	.326
100		.946		.630	–	–	.689	.659
95		–		–	.029	.100	–	.015
PGM	(39)	(45)	(68)	(23)	(17)	(50)	(37)	(65)
106	.167	.289	.154	.348	.118	.680	.824	.038
100	.833	.711	.846	.652	.882	.320	.176	.962
97	–	–	–	–	–	–	–	–

Locus/ Allele	UK IWA	UK KHB	UK KNA	UK LLO	UK MFU	IRE MIR	UK RAV	UK USX
AAT–1	(48)	(42)	(79)	(65)	(51)	(51)	(102)	(55)
104	–	–	–	–	–	–	–	–
100	1.000	1.000	1.000	1.000	1.000	1.000	1.000	1.000
97	–	–	–	–	–	–	–	–
94	–	–	–	–	–	–	–	–
GPI	(48)	(42)	(79)	(65)	(51)	(51)	(102)	(55)
105	–	–	–	–	–	–	–	–
100	.448	.726	.323	.338	.451	.608	.348	.764
98	–	–	–	–	–	–	–	–
96	.552	.274	.677	.662	.549	.392	.652	.236
93	–	–	–	–	–	–	–	–
90	–	–	–	–	–	–	–	–
IDH–1	(48)	(42)	(79)	(65)	(51)	(51)	(102)	(55)
110	–	–	–	–	–	–	–	–
107	–	–	–	–	–	–	–	–
100	1.000	1.000	1.000	1.000	1.000	1.000	1.000	.945
96	–	–	–	–	–	–	–	.055
91	–	–	–	–	–	–	–	–
IDH–2						(51)		(55)
A						–		–
102						–		–
100						1.000		1.000

Locus/ Allele	UK IWA	UK KHB	UK KNA	UK LLO	UK MFU	IRE MIR	UK RAV	UK USX
MDH–1	(48)	(42)	(79)	(65)	(51)	(51)	(102)	(51)
108	–	–	–	–	–	–	–	–
100	1.000	1.000	1.000	1.000	1.000	1.000	1.000	1.000
94	–	–	–	–	–	–	–	–
92	–	–	–	–	–	–	–	–
MPI	(48)	(42)	(79)	(65)	(51)		(102)	
123	–	–	–	–	–		–	
119	–	–	–	–	–		–	
110	–	–	–	–	–		–	
106	–	–	–	–	–		–	
100	1.000	1.000	1.000	1.000	1.000		1.000	
PEPGL	(47)	(41)	(75)	(64)	(51)		(16)	
113	–	–	–	–	–		–	
109	.383	.134	.220	.250	–		.250	
100	.617	.866	.780	.750	1.000		.750	
95	–	–	–	–	–		–	
PGM	(47)	(42)	(78)	(65)	(51)	(51)	(99)	(55)
106	.383	.345	.224	.031	–	.147	.263	.145
100	.617	.655	.776	.969	1.000	.853	.737	.855
97	–	–	–	–	–	–	–	–

APPENDIX 2B

Allele frequencies for collections of *Armadillidium vulgare* from southern Europe. Refer to Appendix 1 for explanation of population codes. Sample sizes, in number of individuals, are in parentheses.

	Country/population code:							
Locus/ Allele	PG ADS	GR. ALO	GR ATT	IT COR	GR EVI	SP FDO	GR GRC	SP HOS
AAT–1	(14)	(76)	(130)	(33)	(113)	(22)	(18)	(71)
104	.107	–	–	–	–	–	–	.049
100	.893	1.000	1.000	1.000	1.000	1.000	1.000	.951
97	–	–	–	–	–	–	–	–
94	–	–	–	–	–	–	–	–
GPI	(14)	(78)	(130)	(33)	(113)	(22)	(18)	(71)
105	–	–	.031	–	–	–	–	–
100	.893	.994	.850	.061	1.000	1.000	.667	1.000
98	–	–	–	–	–	–	–	–
96	.107	.006	–	.803	–	–	–	–
93	–	–	–	.015	–	–	–	–
90	–	–	.119	.121	–	–	.333	–
IDH–1	(14)	(76)	(130)	(33)	(113)	(22)	(18)	(71)
110	–	.007	–	–	–	–	–	–
107	.107	.007	–	–	–	–	–	.014
100	.750	.974	1.000	.985	1.000	.909	1.000	.859
96	.143	.013	–	.015	–	.091	–	.127
91	–	–	–	–	–	–	–	–

Locus/ Allele	PG ADS	GR ALO	GR ATT	IT COR	GR EVI	SP FDO	GR GRC	SP HOS
IDH–2				(33)				
A				–				
102				–				
100				1.000				
MDH–1	(14)	(76)	(130)	(33)	(113)	(22)	(18)	(71)
108	–	–	–	–	–	–	–	–
100	.929	.954	.531	.712	.996	.773	.139	1.000
94	–	.013	–	–	–	–	–	–
92	.071	.033	.469	.288	.004	.227	.861	–
MPI	(14)	(76)	(108)		(113)	(22)	(18)	(71)
123	–	–	.019		–	–	–	–
119	–	–	.046		–	–	–	.007
110	–	–	.778		.867	.727	.944	.042
106	–	–	.009		.133	–	.056	–
100	1.000	1.000	.148		–	.273	–	.951
PEPGL	(14)	(76)	(130)	(33)	(113)	(21)	(18)	(68)
113	–	–	–	.015	–	–	–	–
109	.500	.375	.269	.015	.823	.238	.306	.368
100	.500	.618	.554	.970	.177	.762	.694	.610
95	–	.007	.177	–	–	–	–	.022
PGM	(14)	(77)	(123)	(33)	(108)	(22)	(18)	(71)
106	.107	–	–	.712	.005	.114	–	.162
100	.893	1.000	1.000	.045	.644	.886	1.000	.838
97	–	–	–	.242	.352	–	–	–

Locus/ Allele	SP JCA	GR KOM	GR LER	GR PAT	PG PDM	IT PNM	SP TEN	FR TFR
AAT–1	(49)	(111)	(85)	(20)	(36)	(12)	(8)	(52)
104	.020	–	–	–	.194	–	–	.010
100	.980	1.000	1.000	1.000	.806	1.000	1.000	.990
97	–	–	–	–	–	–	–	–
94	–	–	–	–	–	–	–	–
GPI	(49)	(110)	(84)	(20)	(36)	(12)	(8)	(52)
105	–	.023	–	–	–	–	–	–
100	1.000	.968	1.000	1.000	1.000	.583	1.000	.894
98	–	–	–	–	–	–	–	–
96	–	–	–	–	–	.417	–	.106
93	–	–	–	–	–	–	–	–
90	–	.009	–	–	–	–	–	–
IDH–1	(49)	(111)	(85)	(20)	(36)	(12)	(8)	(52)
110	–	–	–	–	–	–	–	–
107	–	–	–	–	–	–	–	.038
100	.949	1.000	.259	.625	1.000	1.000	1.000	.962
96	.051	–	.735	.375	–	–	–	–
91	–	–	.006	–	–	–	–	–
IDH–2						(12)		(52)
A						–		–
102						–		–
100						1.000		1.000

Locus/ Allele	SP JCA	GR KOM	GR LER	GR PAT	PG PDM	IT PNM	SP TEN	FR TFR
MDH–1	(49)	(111)	(85)	(20)	(36)	(12)	(8)	(52)
108	–	–	–	–	–	–	–	–
100	.837	1.000	.835	.600	1.000	–	.313	.971
94	–	–	.059	.325	–	–	–	–
92	.163	–	.106	.075	–	1.000	.688	.029
MPI	(49)	(111)	(84)	(15)	(36)		(8)	
123	–	–	–	–	–		–	
119	.102	–	–	.033	–		.063	
110	.490	1.000	.923	.733	–		.563	
106	–	–	.077	.233	–		–	
100	.408	–	–	–	1.000		.375	
PEPGL	(49)	(111)	(85)	(20)	(36)	(12)	(8)	(55)
113	–	–	.012	–	–	–	–	–
109	.194	.378	.976	1.000	.111	–	.313	.473
100	.806	.622	.012	–	.889	1.000	.688	.527
95	–	–	–	–	–	–	–	–
PGM	(49)	(97)	(79)	(15)	(36)	(12)	(8)	(52)
106	.255	.124	.038	–	–	.500	.125	.096
100	.745	.876	.962	1.000	.972	.500	.875	.904
97	–	–	–	–	.028	–	–	–

Locus/ Allele	GR TIN	IT VEN
AAT–1	(105)	(42)
104	–	–
100	1.000	1.000
97	–	–
94	–	–
GPI	(105)	(42)
105	–	–
100	.976	1.000
98	–	–
96	–	–
93	–	–
90	.024	–
IDH–1	(105)	(42)
110	.005	–
107	–	–
100	.995	1.000
96	–	–
91	–	–
IDH–2		(42)
A		–
102		–
100		1.000
MDH–1	(105)	(42)
108	.005	–
100	.995	1.000

Locus/ Allele	GR TIN	IT VEN
94	–	–
92	–	–
MPI	(105)	
123	–	
119	–	
110	1.000	
106	–	
100	–	
PEPGL	(105)	(41)
113	–	–
109	.705	.866
100	.295	.134
95	–	–
PGM	(105)	(42)
106	.148	.036
100	.852	.964
97	–	–

APPENDIX 2C

Allele frequencies for collections of *Armadillidium vulgare* from northeastern North America. Refer to Appendix 1 for explanation of population codes. Sample sizes, in number of individuals, are in parentheses.

	Country or state/population code:							
Locus/	IA	VA	CAN	MI	NY	NY	KY	KY
Allele	AIA	ALX	BCA	BCM	BNY	CPW	EHC	FOA
AAT–1	(98)		(72)	(58)	(43)	(51)		
104	–		–	–	–	–		
100	1.000		1.000	.991	.965	1.000		
97	–		–	.009	.035	–		
94	–		–	–	–	–		
GPI	(98)	(100)	(72)	(58)	(43)	(51)	(43)	(48)
105	–	–	–	–	–	–	–	–
100	.903	.665	.819	.733	.651	.804	.802	.813
98	–	–	–	–	–	–	–	–
96	.097	.335	.181	.267	.349	.196	.198	.188
93	–	–	–	–	–	–	–	–
90	–	–	–	–	–	–	–	–
IDH–1	(98)	(100)	(72)	(58)	(43)	(51)	(43)	(48)
110	–	–	.021	.060	–	–	–	–
107	–	–	–	–	–	–	–	–
100	1.000	1.000	.979	.940	1.000	1.000	1.000	1.000
96	–	–	–	–	–	–	–	–
91	–	–	–	–	–	–	–	–

Locus/ Allele	IA AIA	VA ALX	CAN BCA	MI BCM	NY BNY	NY CPW	KY EHC	KY FOA
IDH–2	(98)	(100)	(72)	(58)	(43)		(43)	(48)
A	–	–	–	–	–		–	–
102	–	–	–	–	–		–	–
100	1.000	1.000	1.000	1.000	1.000		1.000	1.000
MDH–1	(98)	(100)	(72)	(58)	(43)	(51)	(43)	(48)
108	–	–	–	–	–	–	–	–
100	1.000	1.000	.931	.983	.802	1.000	.965	.542
94	–	–	–	–	–	–	–	–
92	–	–	.069	.017	.198	–	.035	.458
MPI						(51)		
123						–		
119						.010		
110						.059		
106						–		
100						.931		
PEPGL						(51)		
113						–		
109						.382		
100						.618		
95						–		
PGM	(98)	(100)	(72)	(58)	(43)	(51)	(43)	(48)
106	.337	.195	.076	.172	.256	.059	.047	.052
100	.663	.805	.924	.828	.744	.941	.953	.948
97	–	–	–	–	–	–	–	–

Locus/ Allele	NY INY	KY KKM	CAN MID	IL MIL	MO MTH	MI NBM	NB OMA	OH RSI
AAT–1	(73)		(93)	(9)	(97)	(70)	(75)	(50)
104	–		–	–	–	–	–	–
100	1.000		1.000	1.000	.995	1.000	1.000	1.000
97	–		–	–	.005	–	–	–
94	–		–	–	–	–	–	–
GPI	(73)	(100)	(93)	(18)	(97)	(70)	(75)	(50)
105	–	–	–	–	–	–	–	–
100	.836	.820	.925	.833	.789	.793	.747	.740
98	–	–	–	–	–	–	–	–
96	.164	.180	.075	.167	.211	.207	.253	.260
93	–	–	–	–	–	–	–	–
90	–	–	–	–	–	–	–	–
IDH–1	(73)	(100)	(92)	(9)	(97)	(70)	(75)	(50)
110	–	–	–	.056	–	.007	.087	–
107	–	–	.212	–	–	.014	–	.020
100	1.000	1.000	.788	.944	1.000	.979	.913	.980
96	–	–	–	–	–	–	–	–
91	–	–	–	–	–	–	–	–

Locus/ Allele	NY INY	KY KKM	CAN MID	IL MIL	MO MTH	MI NBM	NB OMA	OH RSI
IDH–2		(100)		(9)	(97)	(68)	(75)	
A		–		–	–	–	–	
102		–		–	–	.007	–	
100		1.000		1.000	1.000	.993	1.000	
MDH–1	(73)	(100)	(93)	(9)	(97)	(70)	(75)	(50)
108	–	–	–	–	–	–	–	–
100	1.000	.750	1.000	1.000	.995	.907	.820	.990
94	–	–	–	–	–	–	–	–
92	–	.250	–	–	.005	.093	.180	.010
MPI	(73)		(93)					(50)
123	–		–					–
119	–		–					–
110	–		–					–
106	–		–					–
100	1.000		1.000					1.000
PEPGL	(73)		(93)					(50)
113	–		–					–
109	.384		.516					.550
100	.616		.484					.450
95	–		–					–
PGM	(70)	(100)	(86)	(9)	(97)	(70)	(75)	(48)
106	.500	.065	.122	–	.196	.171	.100	.063
100	.500	.935	.878	1.000	.804	.829	.900	.938
97	–	–	–	–	–	–	–	–

Locus/ Allele	KY SBF	MO SCF	MO SLM	MO SMO	KY STR	MI UMC	VA VB2	MA WH
AAT–1			(95)	(63)	(100)	(101)		(85)
104			–	–	–	–		–
100			1.000	.968	1.000	1.000		1.000
97			–	.032	–	–		–
94			–	–	–	–		–
GPI	(100)	(25)	(101)	(63)	(100)	(101)	(51)	(85)
105	–	–	–	–	–	–	–	–
100	.800	.820	.827	.913	.850	.609	.647	.547
98	–	–	–	–	–	–	–	–
96	.200	.180	.173	.087	.150	.391	.353	.453
93	–	–	–	–	–	–	–	–
90	–	–	–	–	–	–	–	–
IDH–1	(50)	(25)	(101)	(63)	(100)	(101)	(51)	(85)
110	–	–	–	–	–	–	.088	–
107	–	–	–	.016	–	–	–	–
100	1.000	1.000	1.000	.984	1.000	1.000	.912	1.000
96	–	–	–	–	–	–	–	–
91	–	–	–	–	–	–	–	–

Locus/ Allele	KY SBF	MO SCF	MO SLM	MO SMO	KY STR	MI UMC	VA VB2	MA WH
IDH–2	(100)	(25)	(101)	(63)	(100)		(51)	(85)
A	–	–	–	–	–		–	–
102	.005	–	–	–	–		–	–
100	.995	1.000	1.000	1.000	1.000		1.000	1.000
MDH–1	(100)	(25)	(101)	(63)	(100)	(101)	(51)	(85)
108	–	–	–	–	–	–	–	–
100	.585	1.000	1.000	.746	.780	.995	.922	1.000
94	–	–	–	–	–	–	–	–
92	.415	–	–	.254	.220	.005	.078	–
MPI						(100)		
123						–		
119						–		
110						.040		
106						–		
100						.960		
PEPGL						(99)		
113						–		
109						.460		
100						.540		
95						–		
PGM	(50)	(25)	(101)	(63)	(100)	(101)	(51)	(85)
106	.120	.300	.168	.167	.060	.104	.137	.094
100	.880	.700	.832	.833	.940	.896	.863	.906
97	–	–	–	–	–	–	–	–

Locus/ Allele	OH WSI	MI WSU
AAT–1	(31)	(35)
104	–	–
100	1.000	1.000
97	–	–
94	–	–
GPI	(31)	(35)
105	–	–
100	.548	.286
98	–	–
96	.452	.714
93	–	–
90	–	–
IDH–1	(31)	(35)
110	–	–
107	.016	–
100	.984	1.000
96	–	–
91	–	–

Locus/ Allele	OH WSI	MI WSU
IDH–2		
A		
102		
100		
MDH–1	(31)	(35)
108	–	–
100	1.000	1.000
94	–	–
92	–	–
MPI	(31)	(35)
123	–	–
119	–	–
110	–	.029
106	–	–
100	1.000	.971
PEPGL	(31)	(34)
113	–	–
109	.484	.574
100	.516	.426
95	–	–
PGM	(31)	(35)
106	.145	.057
100	.855	.943
97	–	–

APPENDIX 2D

Allele frequencies for collections of Armadillidium vulgare from southeastern North America. Refer to Appendix 1 for explanation of population codes. Sample sizes, in number of individuals, are in parentheses.

	Country or state/population code:							
Locus/ Allele	AL ADI	GA AGA	AR ARK	TN CHW	TX DFW	LA ENO	FL FL1	FL FL2
AAT–1	(22)	(52)		(20)	(101)	(47)	(112)	(70)
104	–	–		–	–	–	.022	–
100	.886	1.000		1.000	.985	.894	.973	1.000
97	.114	–		–	.015	.106	.004	–
94	–	–		–	–	–	–	–
GPI	(25)	(52)	(51)	(20)	(101)	(47)	(112)	(70)
105	–	–	–	–	–	–	–	–
100	1.000	.779	.892	.900	.985	1.000	.920	.979
98	–	–	–	–	–	–	–	–
96	–	.221	.108	.100	.015	–	.076	.021
93	–	–	–	–	–	–	.004	–
90	–	–	–	–	–	–	–	–

Locus/ Allele	AL ADI	GA AGA	AR ARK	TN CHW	TX DFW	LA ENO	FL FL1	FL FL2
IDH–1	(25)	(52)	(51)	(20)	(100)	(47)	(112)	(70)
110	–	–	–	–	–	–	–	–
107	–	–	.020	.025	.020	.202	–	.007
100	1.000	.990	.951	.975	.915	.777	1.000	.993
96	–	.010	.029	–	.065	.021	–	–
91	–	–	–	–	–	–	–	–
IDH–2		(52)	(51)	(20)	(101)	(39)		
A		–	–	–	–	.064		
102		–	.176	–	–	–		
100		1.000	.824	1.000	1.000	.936		
MDH–1	(25)	(52)	(51)	(20)	(101)	(47)	(112)	(70)
108	–	–	–	–	–	–	–	–
100	.880	.952	.971	1.000	1.000	.989	.866	.829
94	–	–	–	–	–	–	–	–
92	.120	.048	.029	–	–	.011	.134	.171
MPI	(25)					(47)	(112)	(70)
123	–					–	–	–
119	–					–	–	–
110	.200					.064	.313	.386
106	–					–	–	–
100	.800					.936	.688	.614
PEPGL	(25)					(47)	(58)	(16)
113	–					–	–	–
109	.060					.415	.172	.219
100	.940					.585	.828	.781
95	–					–	–	–
PGM	(25)	(52)	(51)	(20)	(101)	(47)	(112)	(70)
106	.200	.192	.275	.250	.332	.564	.156	.414
100	.800	.808	.725	.750	.668	.436	.844	.586
97	–	–	–	–	–	–	–	–

Locus/ Allele	FL FL3	FL FL4	FL FL5	FL FL6	FL FL7	FL FL8	FL FL9	FL GAI
AAT–1	(60)	(70)	(98)	(42)	(100)	(61)	(50)	(64)
104	–	.036	–	–	–	.008	–	–
100	.992	.957	.995	1.000	1.000	.975	1.000	1.000
97	.008	.007	.005	–	–	.016	–	–
94	–	–	–	–	–	–	–	–
GPI	(60)	(70)	(98)	(42)	(100)	(61)	(50)	(64)
105	–	–	–	–	–	–	–	–
100	.942	.921	.969	.976	.815	.951	.880	.977
98	.008	–	.015	–	–	.033	–	–
96	.050	.079	.015	.024	.185	.016	.120	.023
93	–	–	–	–	–	–	–	–
90	–	–	–	–	–	–	–	–

Locus/ Allele	FL FL3	FL FL4	FL FL5	FL FL6	FL FL7	FL FL8	FL FL9	FL GAI
IDH–1	(60)	(70)	(98)	(42)	(100)	(61)	(50)	(64)
110	–	–	–	–	–	.008	–	–
107	.033	.021	–	.012	–	.008	–	.047
100	.967	.971	1.000	.988	1.000	.984	1.000	.953
96	–	.007	–	–	–	–	–	–
91	–	–	–	–	–	–	–	–
IDH–2								(64)
A								–
102								.047
100								.953
MDH–1	(60)	(70)	(98)	(42)	(100)	(61)	(50)	(64)
108	–	–	–	–	–	–	–	–
100	.858	.879	.944	.976	.930	.951	.920	.898
94	–	–	–	–	–	–	–	–
92	.142	.121	.056	.024	.070	.049	.080	.102
MPI	(60)	(70)	(98)	(42)	(100)	(54)	(47)	
123	–	–	–	–	–	–	–	
119	–	–	–	–	–	–	–	
110	.233	.321	.240	.262	.135	.148	.447	
106	–	–	–	–	–	–	–	
100	.767	.679	.760	.738	.865	.852	.553	
PEPGL	(60)	(70)	(98)	(41)	(99)	(56)	(50)	(64)
113	–	–	–	–	–	–	–	–
109	.258	.236	.143	.463	.399	.402	–	.461
100	.742	.764	.857	.537	.601	.598	1.000	.539
95	–	–	–	–	–	–	–	–
PGM	(60)	(70)	(98)	(42)	(100)	(61)	(50)	(64)
106	.258	.193	.163	.202	.285	.213	.070	.430
100	.742	.807	.837	.798	.715	.787	.930	.570
97	–	–	–	–	–	–	–	–

Locus/ Allele	FL GFL	TN GTN	FL LAN	LA LLA	LA LSU	FL NSB	TN ORT	TN OVT
AAT–1	(98)	(93)	(33)	(89)	(68)	(104)	(62)	(78)
104	–	–	–	–	–	–	–	–
100	1.000	1.000	.833	.983	.978	1.000	1.000	1.000
97	–	–	.167	.017	.022	–	–	–
94	–	–	–	–	–	–	–	–
GPI	(98)	(93)	(33)	(89)	(68)	(104)	(62)	(52)
105	–	–	–	–	–	–	–	–
100	.913	.758	1.000	.966	1.000	.952	.847	.933
98	–	–	–	–	–	–	–	–
96	.087	.242	–	.034	–	.048	.153	.067
93	–	–	–	–	–	–	–	–
90	–	–	–	–	–	–	–	–

Locus/ Allele	FL GFL	TN GTN	FL LAN	LA LLA	LA LSU	FL NSB	TN ORT	TN OVT
IDH–1	(98)	(93)	(33)	(89)	(68)	(104)	(62)	(52)
110	–	.011	–	–	–	–	.016	–
107	–	–	–	.084	.015	–	–	.029
100	1.000	.989	1.000	.899	.985	1.000	.984	.962
96	–	–	–	.017	–	–	–	.010
91	–	–	–	–	–	–	–	–
IDH–2	(98)	(93)	(33)	(89)		(104)	(62)	(78)
A	–	–	–	–		–	–	–
102	.036	–	–	.034		.067	–	–
100	.964	1.000	1.000	.966		.933	1.000	1.000
MDH–1	(98)	(93)	(33)	(89)	(68)	(104)	(62)	(78)
108	–	–	–	–	–	–	–	–
100	.852	1.000	.985	.949	.868	.990	.976	.949
94	–	–	–	–	–	–	–	–
92	.148	–	.015	.051	.132	.010	.024	.051
MPI					(68)			
123					–			
119					–			
110					.257			
106					–			
100					.743			
PEPGL				(25)	(68)	(104)		
113				–	–	–		
109				.600	.154	.308		
100				.400	.846	.692		
95				–	–	–		
PGM	(98)	(90)	(33)	(87)	(68)	(104)	(62)	(52)
106	.270	.306	.364	.218	.272	.284	.226	.413
100	.730	.694	.636	.782	.728	.716	.774	.587
97	–	–	–	–	–	–	–	–

Locus/ Allele	MS SMS	TX TEX	OK TOK	TX WTX
AAT–1	(75)	(52)	(92)	(98)
104	–	.010	–	.031
100	1.000	.990	.989	.934
97	–	–	.011	.036
94	–	–	–	–
GPI	(75)	(52)	(92)	(98)
105	–	–	–	–
100	.947	.981	.962	.964
98	–	–	–	–
96	.053	.019	.038	.036
93	–	–	–	–
90	–	–	–	–
IDH–1	(75)	(52)	(92)	(98)
110	–	–	–	–
107	–	.135	.011	.061

Locus/ Allele	MS SMS	TX TEX	OK TOK	TX WTX
100	1.000	.865	.908	.923
96	–	–	.082	.015
91	–	–	–	–
IDH–2			(92)	(98)
A			–	–
102			.022	.015
100			.978	.985
MDH–1	(75)	(52)	(92)	(98)
108	–	–	–	–
100	.987	.990	.978	.832
94	–	–	–	–
92	.013	.010	.022	.168
MPI	(75)	(52)		
123	–	–		
119	–	–		
110	.107	.221		
106	–	–		
100	.893	.779		
PEPGL	(75)	(51)		
113	–	–		
109	.273	.392		
100	.727	.608		
95	–	–		
PGM	(75)	(52)	(92)	(97)
106	.227	.269	.380	.253
100	.773	.731	.620	.747
97	–	–	–	–

APPENDIX 2E

Allele frequencies for collections of *Armadillidium vulgare* from northwestern North America. Refer to Appendix 1 for explanation of population codes. Sample sizes, in number of individuals, are in parentheses.

	Country or state/population code:							
Locus/	OR	MT	CA	CO	ID	OR	CO	OK
Allele	AOR	CHS	CME	FRU	GFI	GLD	GMC	GOK
AAT–1	(95)	(71)	(104)	(50)		(101)	(29)	(34)
104	.032	–	.067	–		–	–	–
100	.968	1.000	.832	1.000		.980	1.000	1.000
97	–	–	.101	–		.020	–	–
94	–	–	–	–		–	–	–
GPI	(95)	(71)	(104)	(50)	(24)	(101)	(27)	(34)
105	–	–	–	–	–	–	–	–
100	.884	.570	.865	.770	.771	.975	.630	.897
98	–	–	–	–	–	–	–	–

Locus/ Allele	OR AOR	MT CHS	CA CME	CO FRU	ID GFI	OR GLD	CO GMC	OK GOK
96	.116	.430	.135	.230	.229	.025	.370	.103
93	–	–	–	–	–	–	–	–
90	–	–	–	–	–	–	–	–
IDH–1	(95)	(71)	(104)	(50)	(24)	(101)	(29)	(11)
110	–	–	–	–	–	–	.017	–
107	.026	–	.038	–	–	–	–	.045
100	.974	1.000	.962	1.000	1.000	1.000	.983	.955
96	–	–	–	–	–	–	–	–
91	–	–	–	–	–	–	–	–
IDH–2	(95)	(71)	(104)	(50)	(24)	(101)	(29)	(11)
A	–	–	–	–	–	–	–	–
102	–	–	–	–	.021	–	–	–
100	1.000	1.000	1.000	1.000	.979	1.000	1.000	1.000
MDH–1	(95)	(71)	(104)	(50)	(24)	(101)	(28)	(11)
108	–	–	–	–	–	–	–	–
100	.953	1.000	.966	.880	.979	.965	.804	.818
94	–	–	–	–	–	–	–	–
92	.047	–	.034	.120	.021	.035	.196	.182
MPI								
123								
119								
110								
106								
100								
PEPGL					(24)			(34)
113					–			–
109					.521			.235
100					.479			.765
95					–			–
PGM	(95)	(71)	(104)	(50)	(24)	(101)	(29)	(34)
106	.211	.204	.149	.090	.024	.317	.293	.353
100	.789	.796	.851	.910	.958	.683	.707	.647
97	–	–	–	–	–	–	–	–

Locus/ Allele	OK GWO	CA H41	NB INB	CA IRS	ID JID	OR JSA	CO KCO	KS KKN
AAT–1	(65)		(50)	(81)		(92)	(75)	(65)
104	–		–	–		–	–	–
100	1.000		1.000	.975		.995	1.000	1.000
97	–		–	.025		.005	–	–
94	–		–	–		–	–	–
GPI	(65)	(89)	(62)	(81)	(8)	(92)	(75)	(65)
105	–	–	–	–	–	–	–	–
100	.592	.978	.597	.914	.750	.804	.693	.923
98	–	–	–	–	–	–	–	–
96	.408	.022	.403	.086	.250	.196	.307	.077
93	–	–	–	–	–	–	–	–
90	–	–	–	–	–	–	–	–

Locus/ Allele	OK GWO	CA H41	NB INB	CA IRS	ID JID	OR JSA	CO KCO	KS KKN
IDH–1	(65)	(77)	(60)	(81)	(8)	(92)	(75)	(65)
110	–	–	–	–	–	–	–	–
107	–	.123	–	.019	–	–	–	.131
100	.992	.877	1.000	.975	1.000	1.000	1.000	.869
96	–	–	–	.006	–	–	–	–
91	.008	–	–	–	–	–	–	–
IDH–2	(65)	(89)	(60)	(81)	(8)		(75)	(65)
A	–	–	–	–	–		–	–
102	.008	–	–	–	–		–	.031
100	.992	1.000	1.000	1.000	1.000		1.000	.969
MDH–1	(65)	(87)	(60)	(81)	(8)	(92)	(75)	(65)
108	–	–	–	–	–	–	–	–
100	1.000	.759	.808	.840	1.000	.918	.873	.954
94	–	–	–	–	–	–	–	–
92	–	.241	.192	.160	–	.082	.127	.046
MPI						(92)		
123						–		
119						–		
110						.136		
106						–		
100						.864		
PEPGL	(65)		(60)		(8)	(92)		(65)
113	–		–		–	–		–
109	.785		.275		.313	.473		.654
100	.215		.725		.688	.527		.346
95	–		–		–	–		–
PGM	(65)	(89)	(60)	(81)	(8)	(92)	(75)	(65)
106	.554	.107	.067	.235	.125	.022	.367	.085
100	.446	.893	.933	.765	.875	.978	.633	.915
97	–	–	–	–	–	–	–	–

Locus/ Allele	UT MVW	OR OR1	CA PAA	CA SCA	ID SHI	CA SNR	UT SOU	WA SOW
AAT–1	(104)	(38)	(150)	(42)	(75)		(19)	(47)
104	–	–	.023	.095	–		–	.074
100	1.000	1.000	.973	.905	1.000		1.000	.926
97	–	–	.003	–	–		–	–
94	–	–	–	–	–		–	–
GPI	(104)	(38)	(133)	(63)	(75)	(51)	(19)	(47)
105	–	–	–	–	–	–	–	–
100	.668	.908	.917	.944	.707	.824	.632	.691
98	–	–	–	–	–	–	–	–
96	.332	.092	.083	.056	.293	.176	.368	.309
93	–	–	–	–	–	–	–	–
90	–	–	–	–	–	–	–	–
IDH–1	(104)	(38)	(150)	(63)	(75)	(51)	(19)	(47)
110	–	–	–	–	–	–	–	–

Locus/ Allele	UT MVW	OR OR1	CA PAA	CA SCA	ID SHI	CA SNR	UT SOU	WA SOW
107	–	–	.020	.095	–	.069	–	–
100	1.000	1.000	.980	.905	1.000	.931	1.000	1.000
96	–	–	–	–	–	–	–	–
91	–	–	–	–	–	–	–	–
IDH–2	(104)			(63)	(75)	(51)	(19)	
A	–			–	–	–	–	
102	–			–	–	–	.132	
100	1.000			1.000	1.000	1.000	.868	
MDH–1	(104)	(38)	(149)	(63)	(75)	(51)	(19)	(47)
108	–	–	–	–	–	–	–	–
100	.957	.934	.933	.952	1.000	.951	.947	.894
94	–	–	–	–	–	–	–	–
92	.043	.066	.067	.048	–	.049	.053	.106
MPI		(38)	(148)					(47)
123		–	–					–
119		–	–					–
110		.250	.111					.032
106		–	.115					–
100		.750	.774					.968
PEPGL		(38)	(59)					(47)
113		–	–					–
109		.263	.398					.234
100		.737	.602					.766
95		–	–					–
PGM	(104)	(38)	(148)	(63)	(75)	(50)	(19)	(46)
106	.279	.053	.226	.159	.213	.280	.211	.250
100	.721	.947	.706	.841	.787	.720	.789	.750
97	–	–	.068	–	–	–	–	–

Locus/ Allele	CA SRA	CA TIO
AAT–1	(100)	(41)
104	.125	.012
100	.735	.829
97	.140	.159
94	–	–
GPI	(100)	(41)
105	–	–
100	.940	.866
98	–	–
96	.060	.134
93	–	–
90	–	–
IDH–1	(100)	(41)
110	–	–
107	.070	.061

Locus/ Allele	CA SRA	CA TIO
100	.930	.939
96	–	–
91	–	–
IDH–2	(100)	(41)
A	–	–
102	–	–
100	1.000	1.000
MDH–1	(99)	(41)
108	–	–
100	.884	.915
94	–	–
92	.116	.085
MPI		
123		
119		
110		
106		
100		
PEPGL		
113		
109		
100		
95		
PGM	(100)	(41)
106	.175	.305
100	.825	.695
97	–	–

APPENDIX 2F

Allele frequencies for collections of *Armadillidium vulgare* from southwestern North America. Refer to Appendix 1 for explanation of population codes. Sample sizes, in number of individuals, are in parentheses.

	Country or state/population code:							
Locus/ Allele	TX AMA	CA CAT	MEX CBC	CA CGC	AZ CH	TX CTX	CA DEM	TX DRT
AAT–1	(72)	(67)	(102)	(75)	(94)	(61)	(47)	
104	–	–	–	–	–	–	.011	
100	1.000	.993	1.000	1.000	1.000	.984	.989	
97	–	.007	–	–	–	.016	–	
94	–	–	–	–	–	–	–	
GPI	(72)	(67)	(102)	(75)	(104)	(61)	(47)	(72)
105	–	–	–	–	–	–	–	–
100	.903	.933	1.000	1.000	.952	.951	.979	.979
98	–	–	–	–	–	–	–	–

Locus/ Allele	TX AMA	CA CAT	MEX CBC	CA CGC	AZ CH	TX CTX	CA DEM	TX DRT
96	.097	.067	–	–	.048	.049	.021	.021
93	–	–	–	–	–	–	–	–
90	–	–	–	–	–	–	–	–
IDH–1	(72)	(67)	(101)	(75)	(104)	(61)	(47)	(71)
110	–	–	–	–	–	–	–	–
107	.042	.060	–	.007	.019	.107	.106	.197
100	.951	.933	.787	.993	.981	.852	.894	.796
96	.007	.007	.213	–	–	.041	–	.007
91	–	–	–	–	–	–	–	–
IDH–2	(72)	(67)	(102)	(75)	(104)	(61)		(72)
A	–	–	–	–	–	–		–
102	–	.007	–	–	–	–		.014
100	1.000	.993	1.000	1.000	1.000	1.000		.986
MDH–1	(72)	(67)	(102)	(75)	(104)	(61)	(47)	(72)
108	–	–	–	–	–	–	–	–
100	.868	.858	.735	1.000	.928	.943	.894	.965
94	–	–	–	–	–	–	–	–
92	.132	.142	.265	–	.072	.057	.106	.035
MPI							(47)	
123							–	
119							–	
110							.532	
106							–	
100							.468	
PEPGL		(67)		(75)			(47)	
113		–		–			–	
109		.239		.560			.383	
100		.754		.433			.606	
95		.007		.007			.011	
PGM	(72)	(67)	(102)	(75)	(104)	(61)	(47)	(71)
106	.167	.224	.074	.440	.072	.361	.287	.366
100	.833	.776	.926	.560	.928	.639	.713	.634
97	–	–	–	–	–	–	–	–

Locus/ Allele	CA GAV	MEX GMX	TX JTX	AZ KAR	CA LFS	CA LKB	TX LTX	NV LVN
AAT–1	(45)	(36)	(93)	(100)	(104)	(94)	(100)	(60)
104	–	–	–	–	.048	–	–	–
100	1.000	1.000	.952	1.000	.938	.973	.985	.992
97	–	–	.048	–	.014	.027	.015	.008
94	–	–	–	–	–	–	–	–
GPI	(45)	(36)	(93)	(100)	(104)	(94)	(100)	(60)
105	–	–	–	–	–	–	–	–
100	.922	1.000	.957	.965	.909	.984	.855	.908
98	–	–	–	–	–	–	–	–
96	.078	–	.043	.035	.091	.016	.145	.092
93	–	–	–	–	–	–	–	–
90	–	–	–	–	–	–	–	–

Locus/ Allele	CA GAV	MEX GMX	TX JTX	AZ KAR	CA LFS	CA LKB	TX LTX	NV LVN
IDH–1	(45)	(36)	(85)	(100)	(104)	(94)	(99)	(60)
110	–	–	–	–	–	–	–	–
107	.144	–	.059	.040	.250	.080	.071	.058
100	.856	.958	.812	.955	.750	.920	.929	.942
96	–	.042	.129	.005	–	–	–	–
91	–	–	–	–	–	–	–	–
IDH–2	(45)	(36)	(93)	(100)	(104)	(94)	(100)	(59)
A	–	–	–	–	–	–	–	–
102	–	–	.086	–	–	–	.010	–
100	1.000	1.000	.914	1.000	1.000	1.000	.990	1.000
MDH–1	(45)	(36)	(93)	(100)	(104)	(94)	(100)	(60)
108	–	–	–	–	–	–	–	–
100	.756	.736	.957	.970	.913	.883	.950	.833
94	–	–	–	–	–	–	–	–
92	.244	.264	.043	.030	.087	.117	.050	.167
MPI								
123								
119								
110								
106								
100								
PEPGL			(68)		(102)			(54)
113			–		–			–
109			.287		.309			.342
100			.713		.691			.667
95			–		–			.009
PGM	(45)	(36)	(93)	(100)	(104)	(94)	(100)	(50)
106	.144	–	.129	.155	.202	.016	.175	.070
100	.856	1.000	.871	.845	.779	.984	.825	.930
97	–	–	–	–	.019	–	–	–

Locus/ Allele	TX MEN	MEX MTM	CA NIC	CA ORO	TX PAS	MEX PBC	CA PCO	NM PON
AAT–1		(104)	(104)	(53)	(52)	(103)	(72)	(59)
104		–	.005	.066	–	–	–	–
100		1.000	.995	.915	1.000	1.000	1.000	1.000
97		–	–	.019	–	–	–	–
94		–	–	–	–	–	–	–
GPI	(100)	(104)	(104)	(53)	(104)	(103)	(71)	(59)
105	–	–	–	–	–	–	–	–
100	.945	1.000	.889	.858	.918	1.000	.993	.780
98	–	–	–	–	–	–	–	–
96	.055	–	.111	.142	.082	–	.007	.220
93	–	–	–	–	–	–	–	–
90	–	–	–	–	–	–	–	–
IDH–1	(100)	(104)	(104)	(53)	(104)	(103)	(72)	(52)
110	–	–	–	–	–	–	–	.019

Locus/ Allele	TX MEN	MEX MTM	CA NIC	CA ORO	TX PAS	MEX PBC	CA PCO	NM PON
107	.025	.019	.380	.085	.010	–	–	.019
100	.810	.894	.620	.915	.986	.782	1.000	.962
96	.165	.087	–	–	.005	.218	–	–
91	–	–	–	–	–	–	–	–
IDH–2	(100)	(104)	(104)		(104)	(103)	(72)	(72)
A	–	–	–		–	–	–	–
102	.040	.005	–		–	–	–	.019
100	.960	.995	1.000		1.000	1.000	1.000	.981
MDH–1	(100)	(104)	(104)	(53)	(104)	(103)	(72)	(59)
108	–	–	–	–	–	–	–	–
100	.965	.615	.755	.868	.856	.568	.896	.873
94	–	–	–	–	–	–	–	–
92	.035	.385	.245	.132	.144	.432	.104	.127
MPI				(53)				
123				–				
119				–				
110				.321				
106				.028				
100				.651				
PEPGL		(95)						(59)
113		–						–
109		.158						.195
100		.842						.805
95		–						–
PGM	(100)	(78)	(104)	(53)	(104)	(103)	(70)	(59)
106	.315	.077	.101	.160	.048	.010	.107	.144
100	.685	.923	.889	.840	.952	.990	.893	.856
97	–	–	.010	–	–	–	–	–

Locus/ Allele	AZ PRE	TX PRT	CA SAR	AZ SFC	NM SFE	CA SRI	CA SSB	TX STX
AAT–1	(94)	(29)	(100)	(104)	(116)	(19)	(78)	
104	–	–	.005	–	–	–	.019	
100	1.000	1.000	.980	1.000	1.000	1.000	.942	
97	–	–	.015	–	–	–	.038	
94	–	–	–	–	–	–	–	
GPI	(94)	(29)	(100)	(104)	(116)	(19)	(104)	(27)
105	–	–	–	–	–	–	–	–
100	1.000	1.000	.950	.808	.810	.974	.928	.907
98	–	–	–	–	–	–	–	–
96	–	–	.050	.192	.185	.026	.067	.093
93	–	–	–	–	.004	–	–	–
90	–	–	–	–	–	–	.005	–
IDH–1	(94)	(29)	(100)	(104)	(116)	(19)	(104)	(27)
110	.074	–	–	–	.013	–	–	–
107	–	.069	.145	.024	.043	.026	.365	–
100	.926	.931	.855	.976	.940	.947	.635	1.000
96	–	–	–	–	.004	.026	–	–
91	–	–	–	–	–	–	–	–

Locus/ Allele	AZ PRE	TX PRT	CA SAR	AZ SFC	NM SFE	CA SRI	CA SSB	TX STX
IDH–2	(94)	(29)	(100)	(104)		(19)	(104)	(27)
A	–	–	–	–		–	–	–
102	–	–	–	–		–	–	–
100	1.000	1.000	1.000	1.000	1.000	1.000	1.000	
MDH–1	(94)	(29)	(99)	(104)	(73)	(19)	(78)	(27)
108	–	–	–	–	–	–	–	–
100	1.000	.897	.884	.885	.918	.789	.808	.981
94	–	–	–	–	–	–	–	–
92	–	.103	.116	.115	.082	.211	.192	.019
MPI					(116)			
123					–			
119					–			
110					.224			
106					–			
100					.776			
PEPGL		(20)			(55)	(19)		(27)
113		–			–	–		–
109		.575			.309	.158		.537
100		.425			.682	.842		.463
95		–			.009	–		–
PGM	(94)	(28)	(100)	(104)	(116)	(19)	(104)	(27)
106	.207	.500	.235	.096	.211	.158	.091	.185
100	.793	.500	.765	.904	.789	.842	.909	.815
97	–	–	–	–	–	–	–	–

Locus/ Allele	AZ TZ1	AZ TZ4	CA VZ
AAT–1	(104)	(46)	(101)
104	–	–	–
100	1.000	1.000	1.000
97	–	–	–
94	–	–	–
GPI	(104)	(46)	(101)
105	–	–	–
100	.976	.946	.936
98	–	–	–
96	.024	.054	.064
93	–	–	–
90	–	–	–
IDH–1	(104)	(45)	(101)
110	–	–	–
107	.096	.167	.104
100	.889	.833	.886
96	.014	–	.010
91	–	–	–
IDH–2	(104)	(46)	(101)
A	–	–	–

Locus/ Allele	AZ TZI	AZ TZ4	CA VZ
102	.005	–	.005
100	.995	1.000	.995
MDH–1	(104)	(46)	(101)
108	–	–	–
100	.889	.913	.886
94	–	–	–
92	.111	.087	.114
MPI			
123			
119			
110			
106			
100			
PEPGL			
113			
109			
100			
95			
PGM	(104)	(46)	(101)
106	.125	.087	.193
100	.875	.902	.807
97	–	.011	–

REFERENCES

Adamkewicz, S.L. 1969. Colour polymorphism in the land isopod *Armadillidium nasatum. Heredity* 24: 249-264.

Aebersold, P. B., G. A. Winans, D. J. Teel, G. B. Milner, & F. M. Utter 1987. Manual for starch gel electrophoresis: a method for the detection of genetic variation. *NOAA Technical Report NMFS* 61.

Arcangeli, A. 1932. Isopodi terrestri raccolti dal Prof. Silvestri nel Nord-America. *Bollettino del Laboratorio de Zoologia Generale e Agraria, Portici* 26: 121-141.

Baker, A.J. 1992. Genetic and morphometric divergence in ancestral European and descendent New Zealand populations of chaffinches (*Fringilla coelebs*). *Evolution* 46: 1784-1800.

Baker, A.J., & A. Moeed 1987. Rapid genetic differentiation and founder effect in colonizing populations of common mynas (*Acridotheres tristis*). *Evolution* 41: 525-538.

Beck, M. L., and J. O. Price. 1981. Genetic variation in the terrestrial isopod, *Armadillidium vulgare. J. Heredity* 72: 15-18.

Blake, C.H. 1931. Distribution of New England wood lice. *Occ. Papers Boston Soc. Nat. Hist.* 5: 349-355.

Brimley, C.S. 1938. *The insects of North Carolina.* Raleigh: North Carolina Department of Agriculture.

Bryant, E.H., H. Van Dijk, & W. Van Delden 1981. Genetic variability of the face fly, *Musca autumnalis* DeGeer, in relation to a population bottleneck. *Evolution* 35: 872-881.

Chilton, C. 1901. The terrestrial Isopoda of New Zealand. *Trans. Linn. Soc. Lond. (2) Zool.* 8: 99-152.

Christensen, B. 1977. Habitat preference among amylase genotypes in *Asellus aquaticus* (Isopoda, Crustacea). *Hereditas* 87: 21-26.

Clayton, J.W., & D.N. Tretiak 1972. Amine-citrate buffers for pH control in starch gel electrophoresis. *J. Fish. Res. Bd. Canada* 29: 1169-1172.
Cockerell, T.D.A. 1912. The fauna of Boulder County, Colorado, part II. *Univ. Colorado Stud.* 9 (2-3): 41-52.
Cockerell, T.D.A. 1940. The insects of California. *Proc. Sixth Pacific Sci. Cong. Pacific Sci. Assc.* 4: 283-295.
Collinge, W.E. 1915. Some observations on the life history and habits of Isopoda (Woodlice). *Scot. Naturalist* 24: 299-307.
Cooper, D.W. 1968. The significance level in multiple tests made simultaneously. *Heredity* 23: 614-617.
Darwin, C.R. 1859. *On the origin of species by means of natural selection, or, the preservation of favoured races in the struggle for life.* London: John Murray.
DeKay, J.E. 1844. *Zoology of New York, or the New York fauna, part VI, Crustacea.* Albany: Carroll and Cook.
Dollfus, A. 1897. Les Crustacés Isopodes terrestres à grande dispersion. *La Feuille des Jeunes Naturalistes* 324: 205-212.
Drummond, P.C. 1965. The terrestrial isopod crustaceans (Oniscoidea) of Florida. Unpublished MS thesis, University of Florida, Gainesville. Pp. 1-57.
Eberly, W.R. 1953. The terrestrial isopods (Oniscoidea) of Indiana. *Proc. Indiana Acad. Sci.* 63: 272-277.
Enckell, P.H., S. Bengtson, P. Douwes, M. Niklasson, B. Stille, & B. Wiman. 1986. The dispersal pattern of an anthropochorous species: genetic variation in populations of Lumbricus terrestris L. (Lumbricidae) in the Faroe Islands. *Hereditas* 104: 253-261.
Endler, J.A. 1986. *Natural selection in the wild.* Princeton: Princeton University Press.
Essig, E.O. 1926. *Insects of western North America.* New York: Macmillan, New York.
Ferrara, F., & S. Taiti. 1979. A check-list of terrestrial isopods from Africa (south of the Sahara). *Monit. Zool. Ital. n.s., Suppl.* 12: 89-215.
Fitch, A. 1856. The first and second reports on the noxious, beneficial and other insects of New-York. Albany, New York.
Fowler, H.W. 1912. The Crustacea of New Jersey. *Ann. Rept. New Jersey St. Mus.* 1911: 25-650.
Garthwaite R., & C. Sassaman. 1985. *Porcellionides floria*, new species, from North America; provinciality in the cosmopolitan isopod *Porcellionides pruinosus* (Brandt). *J. Crust. Biol.* 5: 539-555.
Gould, A.A. 1841. *Report on the invertebrata of Massachusetts comprising the Mollusca, Crustacea, Annelida, and Radiata.* Cambridge: Folsom, Wells, and Thurston.
Green. A.J.A. 1961. A study of Tasmanian Oniscidea (Crustacea: Isopoda). *Aust. J. Zool.* 9: 258-365.
Haldane, J.B.S. 1962. Evidence for heterosis in woodlice. *J. Genetics* 58: 39-41.
Halpine, T., & S.J. Kerr. 1986. Mutant allele frequencies in the cat population of Omaha, Nebraska. *J. Heredity* 77: 460-462.
Harris, H. & D.A. Hopkinson. 1976. *Handbook of enzyme electrophoresis in human genetics.* New York: American Elsevier.
Hartl, D.L. 1981. *A primer of population genetics.* Sunderland: Sinauer Associates, Sunderland.
Hatch, M.H. 1939. Records of terrestrial Isopoda or sow bugs from North America. *Amer. Midld. Nat.* 21: 256-257.
Hatch, M.H. 1947. The Chelifera and Isopoda of Washington and adjacent regions. *University of Washington Publications in Biology* 10: 155-274.
Hatchett, S.P. 1947. Biology of the Isopoda of Michigan. *Ecol. Monogr.* 17: 47-79.
Heath, D.J., B.J. Riddoch, D. Childs, & J.R. Ratford. 1988. Selection component analysis of the PGI polymorphism in *Sphaeroma rugicauda. Heredity* 60: 229-235.
Howard, H.W. 1943. Length of life of sperm in the woodlouse *Armadillidium vulgare. Nature* 152: 331.
Howard, H.W. 1981. Constancy of polymorphism in three populations of the terrestrial isopod, *Armadillidium vulgare. Heredity* 47: 135-137.

Jackson, H.G. 1941. Check-list of the terrestrial and fresh-water Isopoda of Oceania. *Smith. Misc. Coll.* 99(8): 1-35.

Johnson, C. 1982. Multiple insemination and sperm storage in the isopod, *Venezillo evergladensis* Schultz, 1963. *Crustaceana* 42: 225-232.

Johnson, C. 1985. Selection dynamics in the isopod *Venezillo evergladensis. Pedobiol.* 28: 403-412.

Kimura, M. 1983. *The neutral theory of molecular evolution.* Cambridge: Cambridge University Press.

Koehn, R.K., R.I.E. Newell, & F.W. Immermann. 1980. Maintenance of an aminopeptidase allele frequency cline by natural selection. *Proc. Nat. Acad. Sci. USA* 77: 5385-5389.

Kunkel, B.W. 1918. The Arthrostraca of Connecticut. *Bull. Conn. St. Geol. Nat. Hist. Surv.* 26: 1-261.

Lane, R.L. 1977. A development investigation of the reproductive systems of *Armadillidium vulgare* (Latreille) and *Porcellionides pruinosus* (Brandt) (Isopoda). *Crustaceana* 33: 237-247.

Levene, H. 1949. On a matching problem arising in genetics. *Ann. Math. Stat.* 20: 91-94.

Li, C.C., & D.G. Horvitz. 1953. Some methods of estimating the inbreeding coefficient. *Amer. J. Human Genetics* 5: 107-117.

Lokke, D.H. 1966. Mass movements of terrestrial isopods related to atmospheric circulation patterns. *Trans. Kansas Acad. Sci.* 69: 117-122.

Longnecker, M. 1923. The terrestrial isopods of Iowa. *Proc. Iowa Acad. Sci.* 30: 197-199.

Lueken, W. 1962. Zur spermienspeicherung bei Armadillidien (Isopoda terrestria). *Crustaceana* 5: 27-34.

Merriam, H.G. 1971. Sensitivity of terrestrial isopod populations (*Armadillidium*) to food quality differences. *Canad. J. Zool.* 49: 667-674.

Miller, M.A. 1938. Comparative ecological studies on the terrestrial isopod Crustacea of the San Francisco Bay Region. *Univ. Cal. Publ. Zool.* 43: 113-142.

Morrill, R.B., & N.B. Todd 1978. Mutant allele frequencies in domestic cats of Denver, Colorado. *J. Heredity* 69: 131-134.

Muchmore, W.B. 1957. Some exotic terrestrial isopods (Isopoda: Oniscoidea) from New York State. *J. Wash. Acad. Sci.* 47: 78-83

Mulaik, S B. 1960. Contribucion al conocimiento de los isopodos terrestres de Mexico (Isopoda, Oniscoidea). *Rev. Soc. Mex. Hist. Nat.* 21: 79-292

Mulaik, S., & D. Mulaik 1942. New species and records of American terrestrial isopods. *Bull. Univ. Utah* 32: 1-23.

Mulaik, S., & D. Mulaik 1943. New Texas terrestrial isopods with notes on other species. *Bull. Univ. Utah* 34: 3-15.

Murphy, R.W., J. W. Sites, Jr, D.G. Buth, & C.H. Haufler. 1990. Proteins I: isozyme electrophoresis. In: D. H. Hillis and C. Moritz (eds.), *Molecular systematics*, pp. 45-126. Sunderland: Sinauer Associates.

Nei, M. 1977. F-statistics and analysis of gene diversity in subdivided populations. *Ann. Human Genetics* 41: 225-233.

Nei, M. 1978. Estimation of average heterozygosity and genetic distance from a small number of individuals. – Genetics 89: 583-590.

Nei, M. 1983. Genetic polymorphism and the role of mutation in evolution. – In: M. Nei and R. K. Koehn, eds., Evolution of genes and proteins. Pp. 165-190. Sinauer Associates, Sunderland.

Nei. M., & R.K. Chesser. 1983. Estimation of fixation indices and gene diversities. *Ann. Human Genetics* 47: 253-259.

Paulmier, F.C. 1905. Higher Crustacea of New York City. *Bull. N. Y. St. Mus.* 91: 117-189.

Pierce, W.D. 1907. Notes on the economic importance of sowbugs. *Bull. U.S. Dep. Agricul., Bur. Entomol.* 64: 15-22.

Powers, D.A., G.S. Greaney, & A.R. Place 1979. Physiological correlation between lactate dehydrogenase genotype and haemoglobin function in killifish. *Nature* 277: 240-241.

Rathbun, M. J. 1905. Fauna of New England. 5. List of the Crustacea. *Occ. Pap. Boston Soc. Nat. Hist.* 7: 1-117.

Richardson, H. 1905. A monograph on the isopods of North America. *Bull. US Nat. Mus.* 54: i-liii + 1-727.

Rohlf, F.J., & R.R. Sokal 1981. *Statistical tables.* New York: W. H. Freeman and Company.

Saito, S. 1986. Ecological distribution and species diversity of terrestrial isopods in western part of Gunma, central Japan. *Bull. Biogeogr. Soc. Japan* 41: 45-50.

Sars, G.O. 1899. *An account of the Crustacea of Norway. II. Isopoda.* Bergen: Bergen Museum.

Sassaman, C. 1978a. Dynamics of a lactate dehydrogenase polymorphism in the wood louse *Porcellio scaber* Latr.: evidence for partial assortative mating and heterosis in natural populations. *Genetics* 88: 591-609.

Sassaman, C. 1978b. Mating systems in porcellionid isopods: multiple paternity and sperm mixing in *Porcellio scaber* Latr. *Heredity* 41: 385-397.

Sassaman, C. 1979. Genetics of the malate dehydrogenase isozymes of the isopod *Porcellio scaber. J. Exper. Zool.* 210: 507-513.

Sassaman, C., & R. Garthwaite. 1980. Linkage relationships among six loci in the isopod *Porcellio scaber* Latr. *Genetics* 94: s92-s93.

Sassaman, C., and R. Garthwaite. 1981. Conservation of gene linkage in terrestrial isopods. *Genetics* 97: s93.

Sassaman, C., & R.L. Garthwaite. 1985. Independent assortment of four loci in the isopod, *Porcellio dilatatus. J. Heredity* 76: 470-471.

Say, T. 1818. An account of the Crustacea of the United States. *J. Acad. Nat. Sci. Philadelphia* 1(2), 1818: 393-401.

Schultz, G.A. 1961. Distribution and establishment of a land isopod in North America. *Syst. Zool.* 10: 193-196.

Schultz, G.A. 1965. Terrestrial isopods from caves and mines in Texas and northern Mexico with a description of *Venezillo tanneri* (Mulaik and Mulaik) allotype. *Texas J. Sci.* 17: 101-109.

Schultz, G.A. 1975. Terrestrial isopod crustaceans (Oniscoidea) from coastal sites in Georgia. *Bull. Georgia Acad. Sci.* 34: 185-194.

Schultz, G.A. 1984. *Brackenridgia sphinxensis* n. sp. from a cave with notes on other species from Arizona and California (Isopoda, Oniscoidea). *Southwest. Natural.* 29: 309-319.

Selander, R.K., M.H. Smith, S.Y. Yang, W.E. Johnson, & J.B. Gentry 1971. IV. Biochemical polymorphism and systematics in the genus *Peromyscus.* I. Variation in the old field mouse *(Peromyscus polionotus). Stud. Genetics, IV. Univ. Texas Publ.* 7103: 49-90.

Shihab, A.F., & D.J. Heath 1987. Components of fitness and the PGI polymorphism in the freshwater isopod *Asellus aquaticus* (L). 1. Fecundity selection. *Heredity* 58: 69-73.

Silvestri, F. 1927. Isopodi terrestri raccolti nell'estremo oriente. *Bollettino del Laboratorio de Zoologia Generale e Agraria* 20: 211-269.

Singh, R.S., & A.D. Long 1992. Geographic variation in *Drosophila*: from molecules to morphology and back. *Trends in Ecolo. Evol.* 7:340-345.

Sneath, P.H.A. & R.R. Sokal 1973. *Numerical taxonomy.* San Francisco: W. H. Freeman and Company.

Sokal, R.R. & F.J. Rohlf 1981. *Biometry.* San Francisco: W. H. Freeman and Company.

Sorensen, E.M.B. & R.D. Burkett 1977. A population study of the isopod, *Armadillidium vulgare*, in northeastern Texas. *Southwest. Natural.* 22: 375-388.

Swenk, M.H. 1929. *Armadillidium vulgare* Latr. in Nebraska. *Insect Pest Surv. Bull.* 9: 80.

Swofford, D.L. & R B. Selander 1981. BIOSYS-1: a FORTRAN program for the comprehensive analysis of electrophoretic data in population genetics and systematics. *J. Heredity* 72: 281-283.

Taiti, S. & F. Ferrara. 1991. New species and records of terrestrial isopods (Crustacea) from the Arabian Peninsula. *Fauna of Saudi Arabia* 12: 209-224.

Todd, N.B. 1977. Cats and commerce. *Sci. Amer.* 237: 100-107.

Todd, N.B. & L.M. Todd 1976. Mutant allele frequencies among domestic cats in some eastern areas of Canada. *J. Heredity* 67: 368-372.

Todd, N. B., G. E. Glass, and D. Creel. 1976. Cat population genetics in the U.S. Southwest and Mexico. *Carnivore Genetics Newsletter* 3: 43-54.

Underwood, L.M. 1886. List of the described species of fresh water Crustacea from America, north of Mexico. *Bulle. Ill. St. Lab. Nat. Hist.* 2: 323-386.

Vandel, A. 1941. Recherches sur la génétique et la sexualité des isopodes terrestres. VII. Sur la longevité des spermatozoides à l'intérieur de l'ovaire *d'Armadillidium vulgare. Bull. Biol. Fr. et de la Belg.* 75: 364-367.

Vandel, A. 1962. Isopodes Terrestres (Deuxiéme Partie). *Faune de France* 66: 417-931.

Van Name, W.G. 1936. The American land and fresh-water isopod Crustacea. *Bull. Amer. Mus. Nat. Hist.* 71: i-vii + 1-535.

Van Name, W.G. 1942. A second supplement to the American land and fresh-water isopod Crustacea. *Bull. Amer. Mus. Nat. Hist.* 80: 299-329.

Verspoor, E. 1986. Latitudinal variation of allozyme frequencies in populations of two freshwater isopods in western Europe. *Biochem. Syst. Ecol.* 14: 527-530.

Wahlund, S. 1928. Zusammersetung von populationen und korrelation-sercheinungen von standpunkt der verebungslehre aus betrachtet. *Hereditas* 11: 65-106.

Walker, E.M. 1927. The woodlice or Oniscoidea of Canada (Crustacea, Isopoda). *Canad. Field-Natural.* 41: 173-179.

Warburg, M.R. 1965. The microclimate in the habitats of two isopod species in southern Arizona. *Amer. Midld. Nat.* 73: 363-375.

Watt, W.B., R.C. Cassin, & M.S. Swan 1983. Adaptation at specific loci. III. Field behavior and survivorship differences among Colias PGI genotypes are predictable from *in vitro* biochemistry. *Genetics* 103: 725-739.

Wright, S. 1978. *Evolution and the genetics of populations. 4. Variability within and among natural populations.* Chicago: University of Chicago Press.

Contributors

Alikhan, M. A., Department of Biology, Laurentian University, Sudbury, Ontario P3E 2C6, Canada.

Carefoot, T.H., Department of Zoology, University of British Columbia, 6270 University Blvd. Vancouver, B.C. V6T 1Z4, Canada.

Crawford, C. S., Department of Biology, University of New Mexico, Albuquerque, NM 87131-1091, USA.

Dangerfield, M., Department of Biology, University of Botswana, Private Bag 0022, Gaborone, Botswana.

Drobne, D., Department of Biology, Biotechnical Faculty, University of Ljubljana, Askerceva 12, 61 000 Ljubljana, Slovenia.

Garthwaite, R. L., Osher Lab. Molecular Systematics, California Academy of Sciences, San Francisco, CA 94118-4599, USA.

Heinzelmann, E., Department of Biology, University of New Mexico, Albuquerque, NM 87131-1091, USA.

Hong, G., Department of Biology, Nanjing Normal University, Nanjing 210024, People's Republic of China.

Hornung, E., Department of Ecology, Jozsef Attila University, Pf. 659, Egyetem u. 2, H-6722 Szeged, Hungary.

Lawson, R., Osher Lab. Molecular Systematics, California Academy of Sciences, San Francisco, CA 94118-4599, USA.

Ličar, P., Department of Biology, Biotechnical Faculty, University of Ljubljana, Askerceva 12, 61 000 Ljubljana, Slovenia.

Molles, Jr., M. C., Department of Biology, University of New Mexico, Albuquerque, NM 87131-1091, USA.

O'Donnell, M. J., Department of Biology, McMaster University, 1280 Main Street West, Hamilton, Ontario L8S 4L8, Canada.

Sassaman, C., Department of Biology, University of California, Riverside, CA 92521, USA.

Štrus, J., Department of Biology, Biotechnical Faculty,University of Ljubljana, Askerceva 12, 61 000 Ljubljana, Slovenia.

Szlevecz, K., Soil Zoology Group, Department of Systematic Zoology & Ecology, Eötvös University, Puskin u. 3, H-1088 Budapest, Hungary (Department of Geography & Environmental Engineering, The John Hopkins University, Baltimore MD 21218, USA).

Tang Bo-ping, Department of Biology, Yancheng Normal College, Yancheng 224002, People's Republic of China.

Taylor, B.E., Department of Zoology, University of British Columbia, 6270 University Blvd. Vancouver, B.C. V6T 1Z4, Canada.

Telford, S. R., Department of Zoology, University of Pretoria, Pretoria 0002, South Africa.

Tian, W., Department of Biology, Nanjing Normal University, Nanjing 210024, People's Republic of China.

Toop, W., Department of Zoology, University of Cologne, Weyertal 119, D-50923 Köln, Germany.

Uesbeck M., Department of Zoology, University of Cologne, Weyertal 119, D-50923 Köln, Germany.

Warburg, M. R., Department of Biology, Technion – Israel Institute of Technology, Haifa 32000, Israel.

Weinstein, D., Department of Biology, Technion – Israel Institute of Technology, Haifa 32000, Israel.

Wright, J. C., Department of Biology, McMaster University, 1280 Main Street West, Hamilton, Ontario L8S 4L8, Canada.

Index